$5

Ⓜ
9/23

Environmental Impact of Fertilizer on Soil and Water

ACS SYMPOSIUM SERIES **872**

Environmental Impact of Fertilizer on Soil and Water

William L. Hall, Jr., EDITOR
IMC Global Operations

Wayne P. Robarge, EDITOR
North Carolina State University

Sponsored by the
**Fertilizer and Soils Subdivision of the
ACS Division of Agrochemicals**

American Chemical Society, Washington, DC

Library of Congress Cataloging-in-Publication Data

Environmental impact of fertilizer on soil and water / William L. Hall, Jr., Wayne P. Robarge, editors.

    p. cm.—(ACS symposium series ; 872)

    Developed from a symposium at the 222nd National Meeting of the American Chemical Society, Chicago, Illinois, August 26–30, 2001.

    Includes bibliographical references and index.

    ISBN 0-8412-3811-1

    1. Fertilizers. 2. Fertilizers—Environmental aspects.

    I. Hall, William L., Jr., 1951- II. Robarge, Wayne P. III. Series.

S633.E7   2003
631.8—dc22   2003058329

The paper used in this publication meets the minimum requirements of American National Standard for Information Sciences—Permanence of Paper for Printed Library Materials, ANSI Z39.48–1984.

Copyright © 2004 American Chemical Society

Distributed by Oxford University Press

All Rights Reserved. Reprographic copying beyond that permitted by Sections 107 or 108 of the U.S. Copyright Act is allowed for internal use only, provided that a per-chapter fee of $24.75 plus $0.75 per page is paid to the Copyright Clearance Center, Inc., 222 Rosewood Drive, Danvers, MA 01923, USA. Republication or reproduction for sale of pages in this book is permitted only under license from ACS. Direct these and other permission requests to ACS Copyright Office, Publications Division, 1155 16th St., N.W., Washington, DC 20036.

The citation of trade names and/or names of manufacturers in this publication is not to be construed as an endorsement or as approval by ACS of the commercial products or services referenced herein; nor should the mere reference herein to any drawing, specification, chemical process, or other data be regarded as a license or as a conveyance of any right or permission to the holder, reader, or any other person or corporation, to manufacture, reproduce, use, or sell any patented invention or copyrighted work that may in any way be related thereto. Registered names, trademarks, etc., used in this publication, even without specific indication thereof, are not to be considered unprotected by law.

PRINTED IN THE UNITED STATES OF AMERICA

Foreword

The ACS Symposium Series was first published in 1974 to provide a mechanism for publishing symposia quickly in book form. The purpose of the series is to publish timely, comprehensive books developed from ACS sponsored symposia based on current scientific research. Occasionally, books are developed from symposia sponsored by other organizations when the topic is of keen interest to the chemistry audience.

Before agreeing to publish a book, the proposed table of contents is reviewed for appropriate and comprehensive coverage and for interest to the audience. Some papers may be excluded to better focus the book; others may be added to provide comprehensiveness. When appropriate, overview or introductory chapters are added. Drafts of chapters are peer-reviewed prior to final acceptance or rejection, and manuscripts are prepared in camera-ready format.

As a rule, only original research papers and original review papers are included in the volumes. Verbatim reproductions of previously published papers are not accepted.

ACS Books Department

Contents

Preface..xi

Detection and Prevalence of Perchlorate Ion in Fertilizers

1. Environmental Analysis of Inorganic Anions and Perchlorate by Ion Chromatography..3
   Peter E. Jackson, Dave Thomas, and Kirk Chassaniol

2. Assessment of Perchlorate in Fertilizers..................................16
   Edward Todd Urbansky

3. Perchlorate in Fertilizer? A Product Defense Story................32
   Linda D. Weber, Wayne P. Robarge, William L. Hall, Jr., and David Averitt

4. Reduction of Perchlorate Levels of Sodium and Potassium Nitrates Derived from Natural Caliche Ore..............................45
   A. Lauterbach

Detection and Assessment of Trace Metals in Fertilizers

5. Regulation of Heavy Metals in Fertilizer: The Current State of Analytical Methodology...61
   Peter F. Kane, William L. Hall, Jr., and David W. Averitt

6. Determination of Trace Metal Content of Fertilizer Source Materials Produced in North America......................................75
   Wayne P. Robarge, Dennis Boos, and Charles Proctor

7. Trace Metal Content of Commercial Fertilizers Marketed in Lebanon...90
   Isam Bashour, Ghada Hannoush, and Nasri Kawar

8. Modeling the Distribution of Aluminum Speciation in Soil
   Water Equilibria with the Mineral Phase Jurbanite..........................100
   C. Y. Wang, S. P. Bi, W. Tang, N. Gan, R. Xu, and L. X. Wen

9. Cadmium Accumulation in Wheat and Potato
   from Phosphate and Waste-Derived Zinc Fertilizers.........................112
   W. L. Pan, R. G. Stevens, and K. A. Labno

10. Health Risk Assessment for Metals in Inorganic Fertilizers:
    Development and Use in Risk Management......................................124
    Daniel M. Woltering

Measurement, Impact, and Management of Fertilizer Nutrients

11. Inorganic Nutrient Use in the United States: Past and Present.........151
    W. M. Stewart

12. Documenting Nitrogen Leaching and Runoff Losses
    from Urban Landscapes..161
    J. L. Cisar, J. E. Erickson, G. H. Snyder, J. J. Haydu,
    and J. C. Volin

13. New Tools for the Analysis and Characterization
    of Slow-Release Fertilizers..180
    J. B. Sartain, W. L. Hall, Jr., R. C. Littell, and E. W. Hopwood

14. Impact of High-Yield, Site-Specific Agriculture on Nutrient
    Efficiency and the Environment..196
    Harold F. Reetz, Jr.

15. Assessing the Water Quality Impacts of Phosphorus
    in Runoff from Agricultural Lands..207
    G. Fred Lee and Anne Jones-Lee

16. Fertility Management Effects on Runoff Losses of Phosphorus........220
    H. A. Torbert and K. N. Potter

17. Environmental and Agronomic Fate of Fertilizer Nitrogen...............235
    Robert G. Hoeft

18. Working Together to Make the U.S. Environmental
    Protection Agency Nonpoint Source Program Effective
    and Efficient..244
    Thomas E. Davenport

Indexes

Author Index..257

Subject Index...259

Preface

This book addresses issues related to fertilizer and its impact on the environment. Contributions to this text were derived from a 2001 American Chemical Society Symposium in Chicago, Illinois. Although a number of issues are important to the scientific community relative to fertilizers and environment, the symposium was intended to address only three: Detection and Prevalence of Perchlorate Ion in Fertilizers; Detection and Assessment of Trace Metals in Fertilizers; and Measurement, Impact, and Management of Fertilizer Nutrients. These issues will continue to be debated in many venues for the foreseeable future. The subjects of these debates will include selection of proper analytical methodology, best management practices, assessment of risk, and suitable regulatory actions. Continued research and sound science must be applied to each subject area if we are to maintain the confidence of the public. This goal can be achieved by the combined efforts of U.S. agriculture, the fertilizer industry, and those responsible for assuring public and environmental safety working together to derive economically sound and environmentally safe practices for fertilizer usage.

The book is divided into three sections: each selected by the symposium organizers on the basis of their value and contribution to the current body of science. Each topic is addressed by the author(s) in his/her own way and from varying and unique perspectives. The authors are to be commended for their work and are thanked for their contribution.

The topics discussed in this text are of paramount interest to many stakeholders in industry, agriculture, state and federal agencies, and the public at large. The information in this text should appeal to, and provide information of value, to individuals with backgrounds in a variety of scientific disciplines who are united in the common goal of the safe and effective use of fertilizers. We hope you agree.

William L. Hall, Jr.
IMC Global Operations
3095 County Road 640 West
Mulberry, FL 33860
wlhall@imcglobal.com (email)
863–428–7161 (telephone)

Wayne P. Robarge
Soil Science Department
North Carolina State University
P.O. Box 7619
Raleigh, NC 27695
wayne_robarge@ncsu.edu (email)

Environmental Impact of Fertilizer on Soil and Water

Detection and Prevalence of Perchlorate Ion in Fertilizers

Chapter 1

Environmental Analysis of Inorganic Anions and Perchlorate by Ion Chromatography

Peter E. Jackson, Dave Thomas, and Kirk Chassaniol

Dionex Corporation, 500 Mercury Drive, Sunnyvale, CA 94088

Ion chromatography (IC) has been approved for the analysis of inorganic anions in environmental waters since the mid-1980s, as described in EPA Method 300.0. Recent advances in instrumentation, columns and detection technology have expanded the scope of IC methods for analytes other than common anions, e.g., disinfection byproduct anions, chromate and perchlorate. In this paper, we review recent developments for the determination of low µg/L levels of anions and perchlorate in environmental samples by IC. The application of EPA Method 314.0 for the analysis of perchlorate in higher ionic strength samples, such as fertilizers, will be also be considered, in addition to the use of electrospray MS detection as a confirmatory technique for anion identification.

Introduction

Ion chromatography can now be considered a well established, mature technique for the analysis of ionic species. A number of standards organizations, including ASTM, AWWA, and ISO, have regulatory methods of analysis based upon IC *(1)*. The technique is applicable to the determination of a wide range of solutes in diverse sample matrices, although the analysis of inorganic anions in environmental waters remains the single most important application of IC *(2)*.

Recent advances in instrumentation, columns and detection technology have expanded the scope of IC beyond the analysis of the common inorganic anions to include solutes such as disinfection byproduct anions, alkali and alkaline earth cations, chromate and perchlorate. There has also been considerable activity regarding new regulations and methods which use IC for environmental water analysis and a number of new regulatory methods based on IC have been published over the last decade.

These new methods tend to be more complex, i.e., use higher capacity columns, alternate detection schemes, and involve more sample preparation, than was required for the analysis of common anions at mg/L levels. A list of the key regulatory IC methods used for environmental analysis is given in Table I. This paper will review general principles and recent advances in the use of IC for the analysis of inorganic anions and perchlorate. Methods for the determination of low µg/L levels of anions and perchlorate in ground and drinking waters, in addition to higher ionic strength samples, will be discussed.

Table I. Key regulatory methods published for the analysis of inorganic ions by ion chromatography.

Method #	Analyte(s)	Date [a]
U.S. EPA 300.0 (A)	F, Cl, NO_2, Br, NO_3, PO_4, SO_4	1983
Standard Methods 4110	F, Cl, NO_2, Br, NO_3, PO_4, SO_4	1992
U.S. EPA 300.0 (B)	Br, ClO_2, ClO_3	1993
ISO 10304-4	Cl, ClO_2, ClO_3	1997
U.S. EPA 300.1 (B)	BrO_3, Br, ClO_2, ClO_3	1997
ASTM D 6581 - 00	BrO_3, Br, ClO_2, ClO_3	2000
U.S. EPA 317.0	BrO_3, Br, ClO_2, ClO_3	2000
U.S. EPA 321.8	BrO_3	1997
ISO 15601	BrO_3	2000
CA DHS Perchlorate	ClO_4	1997
U.S. EPA 9058	ClO_4	1999
U.S. EPA 314.0	ClO_4	1999
ASTM D 2036-97	CN	1997
EPA Method 218.6	Hexavalent chromium (CrO_4)	1991
ISO 10304-3	I, SCN, S_2O_3, SO_3, CrO_4	1997
ISO 14911-1	Li, Na, NH_4, K, Mn, Ca, Mg, Sr, Ba	1998

[a] Date of first publication, earlier methods may have since been revised.

Principles of Ion Chromatography

Ion chromatography is essentially a liquid chromatographic technique applied specifically to the determination of ionic solutes. Ionic species routinely analyzed by IC include, inorganic anions; inorganic cations, including alkali metal, alkaline earth, transition metal and rare earth ions; low molecular weight carboxylic acids plus organic phosphonic and sulfonic acids, including detergents; low molecular weight organic bases; and ionic metal complexes. The instrumentation used for IC is similar to that employed for conventional high performance liquid chromatography (HPLC), although the wetted surfaces of the chromatographic system are typically made of an inert polymer, such as PTFE or (more commonly) PEEK, rather than stainless steel *(3)*. This is due to the fact that the corrosive eluents and regenerant solutions used in IC, such as hydrochloric or sulfuric acids, can contribute to corrosion of stainless steel instrument components. Ion chromatography also differs from HPLC in that ion exchange is the primary separation mode, although other approaches, such as ion exclusion or reversed phase ion pairing, can be used to separate ionic (or ionizable) compounds

The other major difference between IC and HPLC is that conductivity is the primary detection method, as opposed to UV/VIS in conventional HPLC. Conductivity is a bulk property detector and provides universal (non-selective) response for charged, ionic compounds. Conductivity detection can be operated in the direct (or non-suppressed) mode or with the use of an ion exchange-based device, termed a suppressor, which is inserted between the ion exchange column and the conductivity detector. The suppressor is a post-column reaction device unique to IC which greatly improves the signal-to-noise ratio for conductivity detection by reducing the background conductance of the eluent and enhancing the detectability of the eluted ions. In addition to conductivity, other detection methods, such as UV/VIS or amperometry, have proven to be highly sensitive for certain UV absorbing or electroactive species, while post-column derivatization followed by UV/VIS absorption or fluorescence is an important detection approach for selected anions, transition metals, lanthanides and actinides. Also, the use of advanced detection techniques, such as MS and ICP-MS, coupled to IC separations continues to increase *(2)*.

Advances in IC instrumentation have generally kept pace with improvements in conventional HPLC systems. Typical peak area and retention time reproducibility obtained for inorganic analytes at mg/L levels is in the order of 0.5% and 0.2% RSD, respectively *(4)*. Method detection limits (MDLs) are typically in the low µg/L range for most inorganic analytes under standard operating conditions, although they can be significantly lower depending upon the application.

Common Inorganic Anion Analysis

The U.S. National Primary Drinking Water Standards specify a Maximum Contaminant Level (MCL) for a number of common inorganic anions, including fluoride, nitrite and nitrate. The MCLs are specified to minimize potential health effects arising from the ingestion of these anions in drinking water. Consequently, the analysis of these anions in drinking waters is mandated, as are the analytical methods which can be used for their quantification. Other common anions, such as chloride and sulfate, are considered secondary contaminants. The Secondary Drinking Water Standards are guidelines regarding taste, odor, color and certain aesthetic effects which are not federally enforced. However, they are recommended to all the States as reasonable goals and many of the States adopt their own enforceable regulations governing these contaminants (1).

Ion chromatography has been approved for compliance monitoring of these common inorganic anions in drinking water in the U.S. since the mid-1980's, as described in EPA Method 300.0. This method specifies the use of a Dionex IonPac AS4A anion exchange column with an eluent of 1.7 mM sodium bicarbonate / 1.8 mM sodium carbonate for the separation of common anions. An optional column may be substituted provided comparable resolution of peaks is obtained and that the quality control requirements of the method can be met (5). Conductivity is used as a bulk property detector for the measurement of inorganic anions after suppression of the eluent conductance with an Anion MicroMembrane Suppressor (AMMS) operated in the chemical regeneration mode.

Figure 1(A) shows a chromatogram of a standard containing low-mg/L levels of inorganic anions obtained using a recently developed IonPac AS14A anion exchange column with an Anion Self-Regenerating Suppressor (ASRS). The higher capacity AS14A column provides better overall peak resolution compared to the IonPac AS4A column originally specified in Method 300.0, complete resolution of fluoride from acetate, and improved resolution of fluoride from the void peak. All the anions of interest are well resolved within a total run time of less than 10 minutes. The ASRS provides similar method performance to the AMMS originally specified in Method 300.0, but with added convenience as the regenerant solution is electrolytically generated from the conductivity cell effluent. Figure 1(B) shows a chromatogram of a drinking water sample obtained using the IonPac AS14A column and ASRS device. The linear concentration range, coefficients of determination (r^2), and calculated MDLs which can be achieved for each of the anions using Method 300.0 with an AS14A column and ASRS suppressor are shown in Table II.

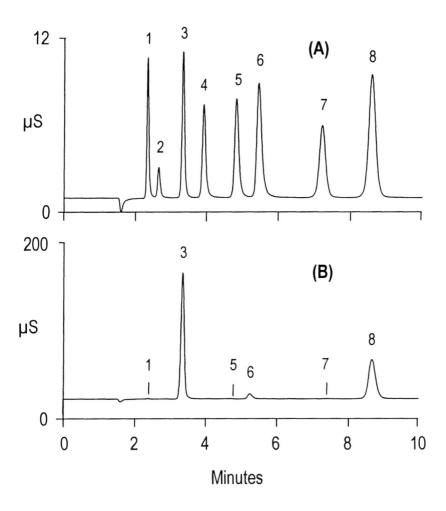

Figure 1. Separation of inorganic anions and acetate using a block-grafted AS14A column. Conditions: column, IonPac AS14A (3 mm ID); eluent, 8.0 mM sodium carbonate / 1.0 mM sodium bicarbonate; flow-rate, 0.5 mL/min; detection, ASRS-ULTRA (2 mm) operated at 50 mA in recycle mode; injection volume, 25 µL; samples, (A) mixed standard, (B), Sunnyvale, CA tapwater; solutes, (A) 1 - fluoride (1 mg/L), 2 - acetate (4 mg/L), 3 - chloride (2 mg/L), 4 - nitrite (3 mg/L), 5 - bromide (5 mg/L), 6 - nitrate (5 mg/L), 7 - phosphate (8 mg/L), 8 - sulfate (6 mg/L); (B) 1 - fluoride (0.03 mg/L), 3 - chloride (31.2 mg/L), 5 - bromide (0.05 mg/L), 6 - nitrate (4.5 mg/L), 4 - phosphate (0.06 mg/L); 5 - sulfate (31.0 mg/L). Chromatograms courtesy of Dionex Corporation.

Table II. U.S. EPA Method 300.0 peformance obtained using an IonPac AS14A column and ASRS suppressor in recycle mode.

Analyte [a]	Range (mg/L)	Linearity (r^2)	MDL [b] ($\mu g/L$)
Fluoride	0.1 - 100	0.9983	3.1
Chloride	0.2 - 200	0.9996	5.4
Nitrite-N	0.1 - 100	0.9999	1.8
Bromide	0.1 - 100	0.9979	8.9
Nitrate-N	0.1 - 100	0.9979	1.7
Phosphate-P	0.1 - 100	0.9981	5.1
Sulfate	0.2 - 200	0.9988	9.6

[a] 150 x 3.0 mm ID column and 25 µL injection.
[b] MDL = (t) x (S), where t = 3.14 for seven replicates, and S = standard deviation of the replicate analyses.

Perchlorate Analysis

Ammonium perchlorate, a key ingredient in solid rocket propellants, has been found in ground and surface waters in a number of States in the U.S., including California, Nevada, Utah, Texas, New York, Maryland, Arkansas, and West Virgina *(6)*. Perchlorate poses a human health concern as it interferes with ability of the thyroid gland to utilize iodine to produce thyroid hormones. Current data from the U.S. EPA indicates that exposure to less than 4-18 µg/L perchlorate provides adequate health protection. Perchlorate contamination of public drinking water wells is a serious problem in California where the Department of Health Services (DHS) has adopted an action level for perchlorate of 18 µg/L *(7)*.

Perchlorate is listed on the U.S. EPA Contaminant Candidate List (CCL) as a research priority under the categories of health, treatment, analytical methods and occurrence priorities *(8)*. In addition, the EPA has recently revised the Unregulated Contaminant Monitoring Rule (UCMR) and added perchlorate to List 1 for assessment monitoring *(8)*. Monitoring of List 1 contaminants has commenced at 2,774 large Public Water Systems (PWS) and a representative sample (800 out of 65,600) of small PWS, as of January 1, 2001 *(8)*. The monitoring results from these systems will be used to estimate the national occurrence of the compounds on List 1 and the data will then be used to evaluate and prioritize contaminants on the CCL.

The determination of trace perchlorate is a difficult analytical task and ion chromatography is perhaps the only viable means for the quantification of such low levels of perchlorate. Large, polarizable anions, such as perchlorate, are strongly retained on conventional anion exchange resins and often display poor peak shape *(6)*. Consequently, the analysis of perchlorate is typically performed using an hydrophilic anion exchange column with an organic modifier, such p-cyanophenol, added to the mobile phase to minimize adsorption and improve peak shape *(9)*. In 1997, in order to support the CA action level of 18 µg/L, the California DHS developed an IC method based upon this separation approach. The CA DHS method used an IonPac AS5 column, with an eluent of 120 mM hydroxide / 2 mM p-cyanophenol, a large loop injection (740 µL) and suppressed conductivity detection using an AMMS in external chemical mode, to achieve an MDL for perchlorate of 0.7 µg/L *(9)*.

It was subsequently shown that an IonPac AS11 column, with an eluent of 100 mM hydroxide, a 1000 µL injection and suppressed conductivity detection using an ASRS in external water mode could achieve an MDL for perchlorate of 0.3 µg/L *(10)*. U.S. EPA Method 9058, published by the EPA Office of Solid Waste and Emergency Response, includes conditions for using either the IonPac AS5 or AS11 columns *(11)*.

In addition, the U.S. EPA Office of Water has promulgated Method 314.0 for the analysis of perchlorate as required by the recent changes to the UCMR. This method is based on the use of a high capacity IonPac AS16 column, large loop injection and suppressed conductivity detection using an ASRS in external water mode *(12)*. The AS16 column is packed with a very hydrophilic anion exchange resin, which allows the elution of the hydrophobic perchlorate ion with good peak shape and high chromatographic efficiency *(6)*. Figure 2(A) shows a chromatogram of a 20 µg/L perchlorate standard obtained using the IonPac AS16 column with a 1000 µL injection loop, a 65 mM hydroxide eluent, and suppressed conductivity detection. Under these conditions, perchlorate elutes within 10 minutes while the common inorganic anions all essentially elute at the column void volume.

Method 314.0 has a calculated MDL of 0.5 µg/L, based upon the standard deviation obtained from seven replicate injections of a 2 µg/L standard. The method is linear in the range of 2.0 to 100 µg/L and quantitative recoveries are obtained for low µg/L levels of perchlorate in spiked drinking and ground water samples *(12)*. A chromatogram of drinking water spiked with perchlorate is shown in Figure 2(B).

Ground water samples may contain high concentrations of common anions, particularly carbonate, chloride, and sulfate. The high capacity AS16 column column can tolerate elevated levels of common inorganic anions, as shown in Figure 3. This shows an overlay of the chromatograms of 20 µg/L perchlorate in

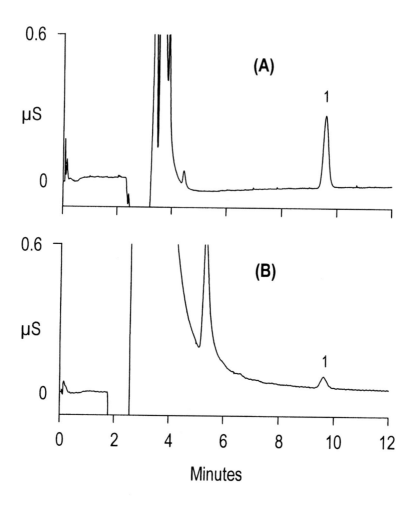

Figure 2. Determination of perchlorate using an AS16 column. Conditions: column, IonPac AS16; eluent, 65 mM potassium hydroxide; eluent source, EG40; flow-rate, 1.2 mL/min; detection, suppressed conductivity with ASRS-ULTRA operated at 300 mA in external water mode; injection volume, 1000 µL; samples, (A) standard, (B), Sunnyvale, CA tap water spiked with 4.0 µg/L perchlorate; solutes, (A) 1 - perchlorate (20 µg/L), (B) 1 - perchlorate (3.8 µg/L). Chromatograms courtesy of Dionex Corporation.

the presence of 0, 50, 200, 600, and 1000 mg/L sulfate. The standard perchlorate retention time decreased only slightly (from 9.55 min to 9.40 min) in the presence of 1000 mg/L sulfate, peak efficiency also decreased slightly (from 6997 theoretical plates (USP) to 6310 theoretical plates) in the presence of 1000 mg/L sulfate, while the peak gaussian factor was unchanged in the presence of up to 1000 mg/L sulfate. These modest changes in retention time and peak efficiency in the presence of up to 1000 mg/L sulfate did not affect the identification or integration of the perchlorate peak. Similar plots were obtained for perchlorate in the presence of 50-1000 mg/L carbonate or chloride. Quantitative recoveries (80-120%) were obtained for 20 µg/L perchlorate spiked in the presence of up to 1000 mg/L sulfate, chloride or carbonate *(6)*.

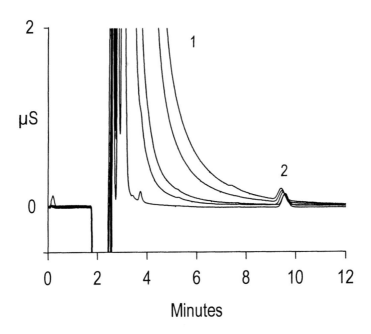

Figure 3. Perchlorate in the presence of 0, 50, 200, 600 and 1000 mg/L of sulfate. Conditions: as for Figure 2 except; eluent, 50 mM potassium hydroxide; flow-rate, 1.5 mL/min; peaks, 1 - sulfate at 0, 50, 200, 600 and 1000 mg/L, 2 - perchlorate (20 µg/L). Chromatogram courtesy of Dionex Corporation.

The analysis of perchlorate in other high ionic strength matrices is also important, e.g., samples such as fermentation media from bioreactors, brines (produced from the regeneration of ion exchange cartridges used in water treatment), and fertilizers. Perchlorate is known to occur naturally in Chilean caliche ores and concern has been raised about potassium nitrate and sodium nitrate fertilizers possibly being another means of introducing perchlorate into the environment *(13)*.

Figure 4 shows a chromatogram of perchlorate in an extract of Chile saltpeter (sodium nitrate). An aqueous extract of the sample (4g/40mL) was diluted 1 to 1000 prior to analysis by IC. Perchlorate, present at 176 µg/L, can be easily quantified in this diluted, aqueous extract which also contains 60 mg/L of nitrate. Further detail on the determination of perchlorate in fertilizer samples can be found in subsequent chapters in this text.

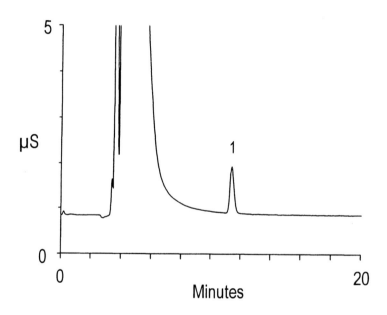

Figure 4. Determination of perchlorate in Chile saltpeter extract. Conditions: as for Figure 2 except; flow-rate, 1.0 mL/min; injection volume, 500 µL; sample, aqueous extract of Chile saltpeter (4g/40mL) diluted 1 to 1000, peaks, 1 - perchlorate (176 µg/L). Chromatogram courtesy of Dionex Corporation.

The use of the approach described in EPA Method 314.0, i.e., large loop injection onto an IonPac AS16 column with an hydroxide eluent and suppressed conductivity detection, is capable of detecting low µg/L levels of perchlorate in high mg/L levels of common anions. However, conductivity is a non-selective detection approach, and so the presence of very high concentrations of matrix ions can make the trace determination of perchlorate difficult, if not impossible, in some sample matrices. In addition, polyphosphate anions, e.g., pyrophosphate and tripolyphosphate, can co-elute with perchlorate depending upon the concentration of the hydroxide eluent *(14)*. Hence, there are occasions when the use of a selective detection approach, such as mass spectrometry, is beneficial for the analysis of perchlorate.

Figure 5(A) shows a chromatogram of an attempt to determine perchlorate in a reclaimed municipal wastewater using IC with suppressed conductivity detection. Although there were several small peaks near the retention time of the target analyte, none of those peaks matched the retention time of perchlorate exactly. Using electrospray ionization mass spectrometry (ESI-MS) as a more selective detection technique, perchlorate could easily be detected and quantitated, at a level of 2.6 µg/L, as illustrated in Figure 5(B).

The two mass chromatograms in Figure 5(B) were obtained by selected ion monitoring at two different *m/z* values, 99 for $^{35}ClO_4^-$ and 101 for $^{37}ClO_4^-$. The areas for the perchlorate peak in the two chromatograms are different, reflecting the natural abundance ratio of the two chlorine isotopes, ^{35}Cl and ^{37}Cl, of about 3:1. Therefore, the ESI-MS analysis provides additional confirmation that the detected solute contains chlorine. Because the mass spectral background and noise at *m/z* 101 are lower than at *m/z* 99, both signals are equally suitable for the quantification of perchlorate. Method detection limits using MS detection were derived by calculating the standard deviation of the results of seven replicate analyses of a low-level standard, as described in Method 314.0 protocol *(10)*. Using the results of seven replicate injections of a 1.0 µg/L standard, an MDL of about 0.3 µg/L was calculated for both signals *(15)*.

Conclusions

Many regulatory and standards organizations, such as the U.S. EPA, ASTM, and ISO, have approved methods of analysis based upon IC, most of which have been published within the last 10 years. These recently developed methods tend to reflect general advances in the field of IC, such as the use of higher capacity columns, large loop injections, and more complex sample preparation and detection schemes. These advances have allowed the

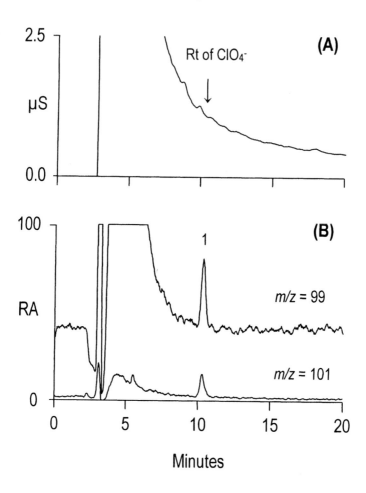

Figure 5. Determination of perchlorate in reclaimed wastewater using IC-MS. Conditions: guard column, IonPac AG16 (2 mm ID); analytical column, IonPac AS16 (2 mm ID); eluent, 65 mM sodium hydroxide; flow-rate, 0.3 mL/min; injection volume, 250 μL; detection, (A) suppressed conductivity with ASRS-ULTRA operated at 300 mA in external water mode, (B) Finnigan AQA MS operated in negative ESI mode, ESI probe at 300°C and -2.5 kV, source CID voltage at 10 V, selected ion monitoring at m/z = 99 and 101, 15-point boxcar smoothing; peaks, 1 - perchlorate (2.6 μg/L). Chromatogram courtesy of Dionex Corporation.

determination of inorganic environmental contaminants at lower detection limits and also expanded the range of analytes which can be measured.

U.S. EPA Method 314.0 was recently developed in order to permit the low μg/L determination of perchlorate in complex matrices. This method utilizes a hydrophilic IonPac AS16 column, hydroxide eluent, large loop injection and suppressed conductivity detection to provide a simple, interference-free method for the determination perchlorate in ground and drinking waters. The MDL of 0.5 μg/L permits quantification of perchlorate below the levels that ensure adequate health protection. The AS16 column provides improved method performance for the analysis of high conductance matrices and permits the determination of perchlorate in high ionic strength samples, such as fertilizers and brines. The use of IC coupled with ESI-MS detection provides postive identification of perchlorate in complex matrices at low μg/L levels.

References

1. Jackson, P.E. *Trends Anal. Chem.* **2001**, 20, 320.
2. Jackson, P.E. In *Encyclopedia of Analytical Chemistry*; Meyers, R.A., Ed.; John Wiley & Sons Ltd.: Chichester, UK, 2000; Vol. 3, pp. 2779.
3. Haddad, P.R.; Jackson, P.E. *Ion Chromatography: Principles and Applications*, J. Chromatogr. Library, Vol. 46; Elsevier: Amsterdam, 1990; pp. 2.
4. Jackson, P.E.; Thomas, D.; Donovan, B.; Pohl, C.A.; Kiser, R.E. *J. Chromatogr. A* **2001**, 920, 51.
5. U.S. EPA Method 300.0, U.S. EPA: Cincinnati, OH, 1993.
6. Jackson, P.E.; Gokhale, S.; Streib, T.; Rohrer, J.S.; Pohl, C.A. *J. Chromatogr. A* **2000**, 888, 151.
7. California Department of Health Services, Update September 5, 2001, http://www.dhs.cahwnet.gov/ps/ddwem/ chemicals/perchl.htm.
8. Pontius, F.W. *J. Am. Water Works Assoc.* **1999**, 91(12), 16.
9. California Department of Health Services, Determination of Perchlorate by Ion Chromatography, 1997.
10. Jackson, P.E.; Laikhtman, M.; Rohrer, J. *J. Chromatogr. A* **1999**, 850, 131.
11. U.S. EPA Method 9058, U.S. EPA: Washington, DC, 1999.
12. U.S. EPA Method 314.0, U.S. EPA: Cincinnati, OH, 1999.
13. U.S. EPA Report EPA/600/R-01/049, U.S. EPA: Cincinnati, OH, 2001.
14. Application Note 144, Dionex Corporation: Sunnyvale, CA, 2001.
15. Roehl, R.; Slingsby, R.; Avdalovic, N.: Jackson, P.E., submitted for publication in *J. Chromatogr. A*.

Chapter 2

Assessment of Perchlorate in Fertilizers

Edward Todd Urbansky

Office of Research and Development, National Risk Management Research Laboratory, Water Supply and Water Resources Division, U.S. Environmental Protection Agency, 26 West Martin Luther King Drive, Cincinnati, OH 45268

Perchlorate has been positively detected only in those materialsknown to be derived from Chilean caliche, which constitute less than 0.2% of U.S. fertilizer application. The data obtained in the preponderance of investigations suggest that fertilizers do not contribute to environmental perchlorate contamination other than in the case of natural saltpeters or their derivatives. Consequently, fertilizers cannot be viewed as major contributors of perchlorate to the environment.

Sources of Perchlorate Contamination

Introduction

Perchlorate (ClO_4^-) was discovered in U.S. waterways in the late 1990s. Most perchlorate salts are used as solid oxidants or energy boosters in rockets or ordnance; therefore, much of the perchlorate-tainted waterways in the U.S. can be traced to military operations, defense contracting, or manufacturing facilities. Perchlorate ion is linked to thyroid dysfunction, due to its similarity in ionic radius to iodide (*1*). Because perchlorate-tainted waters are used for irrigation, there are questions about absorption, elimination, and retention in food plants. Furthermore, recent reports have suggested that fertilizers could represent another source of perchlorate in the environment. Sporadic findings of perchlorate in fertilizers were initially alarming because of the widespread use of fertilizers in production farming.

Because of the dependence of U.S. agriculture on chemical commodity fertilizers, it was clear that assessment of any possible role of fertilizers would require investigation. Attention has been drawn to the possible roles of fertilizers in environmental perchlorate contamination for two reasons. First, perchlorate-tainted agricultural runoff could lead to pollution of natural waterways used as drinking water sources. Second, there is a potential for food plants to take up soluble compounds, including perchlorate salts. There is accordingly a route of human exposure. It has long been known that Chile possesses caliche ores rich in sodium nitrate ($NaNO_3$) that coincidentally are also a natural source of perchlorate (2-3). The origin of the perchlorate anion remains an area of debate, but it is nonetheless present and can be incorporated into any products made from the caliche.

Nutrient Availability

To minimize the need for multiple applications and to prevent overdosing, timed-, delayed-, or controlled-release fertilizers are used in both agricultural and horticultural applications. There are two mechanisms to delay nutrient release. The first is to use essentially insoluble minerals that are not readily converted to absorbable aqueous phase nutrients, for example phosphate rock or other calcium phosphates. The second is to coat the soluble fertilizer with an insoluble material, such as a urea-based polymer or sulfur. This is often done with consumer products, e.g., lawn fertilizers. Most urea-based polymers are methylene ureas or ureamethanal blends. As urea polymers are hydrolyzed, they too serve as a source of nutrients.

Many commodity chemicals used as agricultural fertilizers contain fairly high concentrations of one, or sometimes two, of the primary plant nutrients. A partial list of the major ones includes anhydrous ammonia (NH_3, 82-0-0); ammonium nitrate (NH_4NO_3, 34-0-0); urea [$(NH_2)_2CO$, 46-0-0]; ammonium monohydrogen phosphate [diammonium phosphate, $(NH_4)_2HPO_4$, 18-46-0]; potassium chloride (KCl, 0-0-62); potassium magnesium sulfate (langbeinite, $K_2Mg_2(SO_4)_3$, 0-0-22); triplesuperphosphate [hydrous calcium dihydrogen phosphate $Ca(H_2PO_4)_2 \cdot H_2O$, 0-46-0].

Trace metals (e.g., boron, copper, magnesium) can be applied separately or along with these primary nutrients on a farm site. Fertilizer application in production farming is highly dependent on the crop and the native soil. Crops are influenced by climate, weather, topography, soil type, and other factors that are generally similar within a geographical region; therefore, crops and fertilizer use are also similar within such a region. This is of course unsurprising and consistent with agricultural production of dairy foods, corn, tobacco, wheat, etc. Because all plants require the same primary nutrients, there is some usage to provide these regardless of crop. Local soil conditions also dictate what nutrients should be augmented, and there can be large regional variations.

Unlike agricultural fertilizers which generally are derived from local bulk blending sources due to economic reasons, some consumer products can be distributed over larger geographical regions because of the nature of the market. For example, major manufacturers have a limited number of sites dedicated to blending multiplenutrient formulations. These products are often sold as bagged fertilizers through home-improvement centers, nurseries, florists, horticulturists, and department (or other retail) stores. Unlike agricultural fertilizers, consumer products are usually multi-nutrient formulations. Often, trace metals are sometimes incorporated directly into them. Because the average user will apply only a very small amount of trace metals (or even primary nutrients) relative to a production farm, it is more economical, more practical, and more convenient to use multiple-nutrient formulations. Moreover, consumers typically do not have the wherewithal to disperse careful doses of several single-component fertilizers at appropriate times of the growing season.

Multiple-component fertilizers can be timed (controlled) release or soluble blends. Many multiple-component products are intended for soil amendment to lawns or gardens, e.g., 10-10-10, and other common multiple-macronutrient formulations. Water-soluble blends are used to supply nutrients rapidly to growing plants and are generally applied repeatedly during a growing season (as with each watering), whereas timed-release fertilizers allow water to leach nutrients slowly for release to the soil and plants. They are applied perhaps once or twice a year, e.g., a lawn winterizer. Obviously, soluble and insoluble fertilizers cannot be entirely identical chemically. However, the distinction is essentially irrelevant for agricultural fertilizers, which are applied to fortify particular nutrients. Of course, allowances must be made for the bioavailability of these nutrients. As a general rule, agricultural fertilizers are soluble chemicals. Because fertilizer application on production farms is geographically delimited, there is considerable interest in knowing which commodity chemicals might contain perchlorate and how much. Such information might suggest regions for further investigation. Moreover, it will be important to know what crops might potentially be affected—if any.

Nitrogen Sources

The simplest nitrogen source is anhydrous liquid ammonia. Liquid ammonia is stored in bulk tanks and injected directly into the soil. No fertilizer has a higher nitrogen content. Ammonia is made using the Haber process, which entails heating desiccated nitrogen (separated from liquified air) and hydrogen (usually from methane) in the presence of a catalyst at 500-700 /C. Urea is also a common source of nitrogen. Highly soluble in water, urea hydrolyzes to carbonic acid and ammonia, given time. In addition to its use as a fertilizer, a special feed grade of urea is used to supplement cattle feed.

Nitrate salts are also used as fertilizers. Ammonium nitrate is the primary nitrate salt used in production farming. Most—if not all—ammonium nitrate today is made from atmospheric gases. None of the major nitrogen fertilizer producers [Potash Corporation of Saskatchewan (PCS), Agrium, Coastal, Mississippi Chemical, Kemira Dansmark, and IMC] use natural saltpeters in manufacturing. IMC and PCS do not sell nitrate-based fertilizers, focusing instead on urea, ammonium phosphates, and similar nitrogenous compounds. Consequently, perchlorate contamination is not possible from the raw materials. Ammonium nitrate is prepared from nitric acid and ammonia. The alkali metal saltpeters (sodium and potassium nitrates) are also used as nitrogen sources. Their mineral forms are known as soda niter (nitratine) and potash niter (nitrine), respectively. Chile saltpeter ($NaNO_3$) is mined from caliche ores in the North. The mined rock contains veins rich in sodium nitrate. The ore is crushed and mixed with water to dissolve the soluble salts. The sodium nitrate is then recovered from the leachate. Chile's Sociedad Quimíca y Minera S.A. (SQM) reports annual production of about 992,000 tons of nitrate products. SQM North America sold some 75,000 tons to U.S. farmers in 1998. The company touts its products primarily for cotton, tobacco, and citrus fruits (*4*). No other company sells a product derived from caliche as of this writing; however, Potash Corporation of Saskatchewan does own Chilean caliche mines (*5*).

Because nitrate salts (saltpeters) find use as fertilizers, these natural resources have been mined and refined to produce commercial fertilizers for domestic use or for export. Chilean nitrate fertilizers ($NaNO_3$ and KNO_3) are manufactured by SQM. SQM markets its products in the U.S. under the name Bulldog Soda. Chilean nitrate salts are sold to agricultural operations, chemical suppliers, and consumer oriented companies such as Voluntary Purchasing Groups, Inc., or A.H. Hoffman, Inc., who repackage and resell it as Hi-Yield® or Hoffman® nitrate of soda, respectively. Also, secondary users may incorporate Chilean nitrate salts into watersoluble plant foods, lawn fertilizers, and other retail (specialty) products.

Due to cost and availability, Chilean nitrates are niche fertilizers. SQM markets products to growers of tobacco, citrus fruits, cotton, and some vegetable crops, particularly emphasizing that the products are low in chloride co ntent (4).

As noted above, typical American fertilizer consumption is 54 million tons per year; consequently, most U.S. fertilizers are derived from other raw materials. For example, NH_4NO_3, which is often used for purposes similar to $NaNO_3$, is manufactured from methane, nitrogen, and oxygen. There is no evidence that any ammonium nitrate is derived from Chilean caliche. On account of its low usage, perchlorate from Chilean nitrates cannot represent a significant anthropogenic source of perchlorate nationwide, regardless of the perchlorate content. Recent examination of two manufacturing lots found perchlorate concentrations below 2 mg g–1, i.e., < 0.2% w/w, with some lot-to-lot variability (6). However, in a recent letter to EPA, SQM North America's President Guillermo Farias indicated that SQM had modified its refining process to produce fertilizer containing less than 0.01% perchlorate (<0.1 mg g–1); this corresponds to a reduction of 90-95%. SQM monitors its production stream every 2 hours to verify the perchlorate concentration. Accordingly, previous data on perchlorate content are only applicable in a historical sense rather than being reflective of ongoing fertilizer use.

As Table I indicates, there is limited application of natural saltpeters as fertilizers in the U.S. based simply on total consumption. There just is not enough production of the natural materials. Some states keep detailed records on fertilizer use, especially of chemical commodities used in production farming, but others do not. For example, the Office of Indiana State Chemist is required to keep track of only the top ten fertilizers; even ammonium nitrate is not among the top ten in that state. As a result, it is not easy to discern the potential distribution of minor fertilizers known to contain traces of perchlorate salts. Table II gives the tonnage for a few nitrogen fertilizers for several states. The Corn Belt relies heavily on urea and anhydrous ammonia as nitrogen sources, as shown by Indiana and Ohio consumption of these two chemicals in Table II, while ammonium nitrate finds greater use in tobacco-farming states.

Sodium and potassium nitrates make up a small fraction of the nitrate application in the United States; however, prior to the establishment of nitric acid and ammonia factories, natural saltpeters played significant roles in American agriculture. In addition, ammonium nitrate was manufactured from Chile saltpeter before the industrialized oxidation of ammonia to nitric acid became commonplace in the 1940s. Decades ago, ammonium nitrate was

Table I. Consumption (In Tons) of Nitrate Salts in Regions of the Continental United States for the Year Ending June 30, 1998

Region	NH_4^+	Na^+	K^+	Na^+/K^+
New England	2,469	194	142	0
Mid-Atlantic	33,556	260	12,064	0
South Atlantic	162,035	14,870	17,308	21,762
Midwest	81,585	189	1,496	5
Great Plains	436,371	409	296	55
East South Cent.	489,603	7,786	2,122	914
West South Cent.	338,618	4,192	652	2,851
Rocky Mtn.	254,168	831	9,022	0
Pacific	148,340	10,281	146	0
U.S. Total	1,946,868	39,013	46,100	22,762

NOTE: New England (ME, NH, VT, MA, RI, CT); Mid-Atlantic (NY, NJ, PA, DE, MD, WV); South Atlantic (VA, NC, SC, GA, FL); Midwest (OH, IN, IL, MI, WI); Great Plains (MN, IA, MO, ND, SD, NE, KS); East South Central (KY, TN, AL, MS); West South Central (AR, LA, OK, TX); Rocky Mountain (MT, ID, WY, CO, NM, AZ, UT, NV); Pacific (CA, OR, WA); Total U.S. includes HA, AK, PR. SOURCE: Association of American Plant Food Control Officials/The Fertilizer Institute, Commercial Fertilizers 1998. D.L. Terry and B.J. Kirby, Eds. University of Kentucky: Lexington, KY, 1998.

Table II. Annual Consumption/Application (in Short Tons) of Some Nitrogen Fertilizers for Several States

State	NH_3^a 82-0-0	Urea 46-0-0	NH_4NO_3 34-0-0	KNO_3 14-0-46
Arkansas[b]	1,207	465,737	62,003	N.R.c
Georgia[d]	3,859	15,084	56,215	N.R.c
Indiana[b]	193,347	48,478	N.R.	N.R.
Maryland[c]	1,155	11,614	9,518	N.R.
New Mexico[f]	13,747	34,348	2,725	N.R.
Ohio[b]	86,499	115,180	7,516	N.R.
Texas[g]	142,383	74,235	80,120	1,115
West Virginia[f]	390	4,476	1,283	N.R.

a Anhydrous. b 1998. c N.R. = not reported. d July 1998–June 1999. e July 1998–June 1999. f 1997. g March–August 1998.
SOURCES: Arkansas State Plant Board; Association of American Plant Food Control Officials; Plant Food, Feed, & Grain Division, Georgia Department of Agriculture; Office of Indiana State Chemist and Seed Commissioner; Maryland Department of Agriculture, Office of Plant Industries and Pest Management, State Chemist Section; Feed, Seed, and Fertilizer Bureau, New Mexico Department of Agriculture; Office of the Texas State Chemist; West Virginia Agricultural Statistics Service.

prepared from Chilean sodium nitrate by ion exchange rather than from gaseous reactants. Historical use of ammonium nitrate previous to or in the first half of the 20th century might be linked to contaminated groundwater, and has been attributed to one manufacturing facility in Arizona (7). On the other hand, a recent survey of water supplies for perchlorate was unable to detect perchlorate in nearly all of them, and concluded that perchlorate contamination is generally localized and related to point sources (8). Reliable data on the use of natural saltpeters appears to be unavailable. Most nitrate salts are manufactured by ammonia oxidation, which is a chlorine-free process so that the bulk of nitrate salts must be perchlorate-free. That notwithstanding, there is a possibility for contamination of water supplies through application of products derived from Chile saltpeter.

The lack of information on natural attenuation as well as limited knowledge of hydrogeology makes it difficult to determine where and how such problem sites might be found. For this reason, monitoring for perchlorate under the EPA's Unregulated Contaminant Monitoring Rule for drinking water can be expected to provide some of the most useful information. Meanwhile, it is instructive to consider the processes by which major nutrients are produced so as to evaluate the possibilities for contamination.

Phosphate sources

Phosphate rock is mined in a number of states, including Florida, Idaho, and Montana. Florida's phosphate rock deposits are near the surface (-8 m down) and formed 5-15 million years ago. Natural phosphate rocks usually contain a mixture of the apatite minerals. Fluoroapatite has the empirical formula $Ca_5(PO_4)_3F$. Phosphate is generally applied as ammonium monohydrogen phoshate (DAP), ammonium dihydrogen phosphate (MAP), or hydrous calcium dihydrogen phosphate. The ACS reagent is a white crystalline material; however, agricultural DAP is generally a mixture of gray, brown, and/or black pellets. Some of this color is due to residual calcium minerals (e.g., apatites, gypsum) and some is due to natural organic matter. Most agricultural DAP contains 6-15 mol% $NH_4H_2PO_4$. In the Corn Belt, granular triple superphosphate or GTSP (0-46-0) and DAP, dominate the market. About 82% of the ortho-phosphoric acid produced in the U.S. goes into fertilizer manufacture, with 49% into DAP and 10% into MAP.

Potassium sources

Essentially no true potash (K_2O/K_2CO_3) is used as a fertilizer today, but the name has been retained. Some potassium nitrate and especially potassium chloride (muriate of potash, MOP) dominate this market. Sylvite (KCl) is mined in Canada, and is the most popular potassium source in the Corn Belt. Both Saskatchewan and New Brunswick have sylvite and/or sylvinite mines. The red material with a guaranteed analysis of 0-0-60 contains iron oxide mineral impurties. The pure, clear crystalline material is guaranteed at 0-0-62, which represents the maximum, within the limits of experimental error.

Sylvinite (43% KCl, 57% NaCl) deposits occur in New Mexico and can be refined to remove much of the halite (NaCl). New Mexico also has reserves of sylvite and langbeinite [potassium magnesium sulfate, $K_2Mg_2(SO_4)_3$ or $_2K_2SO_4 \cdot MgSO_4$]. Like Chile saltpeter, these minerals are marine evaporites arising from the drying up of terminal inland seas. In most cases, deposits of sylvite are hundreds or thousands of meters below the surface, having been covered over by sedimentary rock formations over some 300 million years. The market is dominated by two producers, IMC USA and PCS.

Fertilizer Production Recordkeeping

Information on fertilizer production and application comes from a variety of sources, including trade organizations, manufacturers, and government agencies. Both the U.S. Census Bureau and the U.S. Geological Survey track fertilizer commodities. The Census Bureau's Economics and Statistics Administration publishes an annual report (MA325B, formerly MA28B) as well as quarterly reports (MQ28B) on inorganic fertilizer materials and related products. The Geological Survey publishes reports on fertilizer minerals that cover manufacture, use, regulation, litigation, and other matters (*5, 9*). Natural Resources Canada also publishes a minerals yearbook (*10*). Most publications are available online.

Because data are obtained through many sources, it is common for there to be inconsistencies as well as apparent inconsistencies. Apparent inconsistencies sometimes arise from how materials are tracked. Production does not always correlate with consumption. Some reporting systems can omit certain manufacturers and/or product blends. In addition, the tables do not account for normal fluctuations in inventory. In other words, a material produced in one year may be sold or used the following year. Taken together, these factors increase the difficulty in monitoring the application of or tracking transactions involving perchloratecontaining materials.

Initial Investigations of Fertilizer for Perchlorate Occurrence

Aside from the analyses of Chilean caliche, there were no studies to suggest that any other processed fertilizer or raw material might contain perchlorate prior to

1998 when perchlorate was reportedly found in several materials no t derived from Chile saltpeter (*11*). The presence of perchlorate was eventually only confirmed in some products, and not in any of the agricultural fertilizers.Subsequent analyses of different bags (likely different lots) of many of the same brands and grades did not show perchlorate (*12, 13*). The data were not widely applicable because of the choice of products; the same raw materials may be used in a variety of products at a point in time. Additionally, a few major companies are responsible for making a large number of products under several brand names. Furthermore, some companies rely on toll manufacturing so that the products are actually made by another company to meet a specific formulation. Accordingly, an error or contamination associated with one raw material could affect a variety of products without regard to company or application. Since those early days, each subsequent study on fertilizer has attempted to address more issues, and study designs have been continually refined based on what was learned in previous investigations.

The first study brought to light a number of important issues for trace analysis of fertilizers. First, most of the research on determining perchlorate to that time had been focused on either finished potable water or raw source water (*14*). Second, fertilizers are considerably more complicated matrices than dilute water solutions. Third, solid fertilizers are not homogeneous. In fact, some are macroscopically heterogeneous, for example, multi-component formulations used as lawn and garden fertilizers. It is possible to sort out the particles by hand. Thus, representative sampling becomes a key issue. Fourth, the effectiveness of the leaching step must be evaluated. Fifth, the materials must be carefully selected to properly reflect the market of interest, e.g., production farming, lawn treatment, vegetable gardens, houseplant foods.

A separate study assessed interlaboratory corroboration, that is, the ability of different labs to analyze the same sample and get the same result (*15*). Samples of a variety of lawn and garden fertilizers were selected from around the country. It did not account for the sources of the commodity chemicals blended into these products, and it did not link manufacturing lots with lots of raw materials. Even bagged fertilizers from different manufacturing lots may be comprised from some of the same raw materials. Therefore, limitations in choice of products prevent extrapolating the results to large scale fertilization (as in production agriculture). Interlaboratory agreement was generally good, indicating that laboratories were able to determine perchlorate in the fertilizer solutions, despite matrix complexity.

While interlaboratory agreement was good on the liquid solutions, values for leachates derived from different samples of the same lot of material varied substantially in some cases. These products will be referred to here as "bagged fertilizers" even though some are so ld in boxes or jars; this co ntrasts with bulk fertilizers sold by railway car or truck and then used for agricultural purposes (or eventually incorporated into a bagged fertilizer). Such products are usually referred to as specialty fertilizers within the industry. While some products are specifically manufactured so as to assure uniform distribution of macronutrients and micronutrients, it is unclear whether perchlorate contamination would also be homogeneous. Heterogeneity of bagged fertilizer products (especially multiple component products) was therefore demonstrated to be a significant matter. Duplicate samples of solid from bagged fertilizers gave aqueous leachates with considerable differences in measured perchlorate concentration: 5.8 ± 0.5 mg g^{-1} versus 2.8 ± 0.3 mg g^{-1} for one product and 0.98 ± 0.14 mg g^{-1} versus 2.7 ± 0.3 mg 11 g^{-1} for another [values here are the average and estimated standard deviation for 10 results (each of which is an average for one of 10 methods]. Such variation may be attributable to sampling error and does not necessarily reflect an actual difference. When some of the same products were subjected to sampling scheme intended to yield a more representative sam ple, considerably lower intersample variability was observed (*13*). Less striking variation in perchlorate distribution within and among bags of sodium nitrate fertilizer has been seen for sm all grab samples of solids, but can be eliminated by rigorous sampling or choosing larger sample sizes (*6*).

A survey of fertilizers that included a variety of specialty and agricultural products from several states was unable to find any perchlorate (*16*). Samples were leached or dissolved and subjected to complexation electrospray ionization mass spectrometry (cESI-MS) or ion chromatography (IC). The only products that were found to contain any perchlorate were those based on Chile saltpeter. W hile this study was the first to include the same products used on production farms, it did not address the issue of sampling.

Fertilizers are normally sampled by collecting cores through piles using Missouri D tube samplers. These samples are combined, riffled, divided, and analyzed. These practices are standard within the industry and regulatory bodies. Across the nation, state chemists or agriculture departments are obligated to examine fertilizers to verify the manufacturers' reported grades. Sampling practices have evolved to fill those needs. Early studies generally did not take these practices into account, concentrating instead on the analysis of the solid once a grab sample had been collected. Sampling is of course important to obtain representative results. The distribution of perchlorate is not uniform in Chilean sodium nitrate. For example, in two lots with average concentrations of 1.5 and 1.8 mg g^{-1}, individual 10.0-g grab samples ranged from 0.74 to 1.96 mg g^{-1} (*6*).

EPA's Most Expansive Survey of Fertilizers

In an effort to take into account the difficulties of the matrix, the problems with sampling, and the app licability of the results, the EPA entered into a collaboration with The Fertilizer Institute, the International Fertilizer Development Center, the Fertilizer Section of the Office of Indiana State Chemist and Seed Commisioner, the North Carolina State University Department of Soil Science (which serves as part of the North Carolina Agricultural Research Service), and IMC Global for the purpose of conducting a more carefully designed and thorough investigation. It is still important to point out that no single study can say once and for all whether there is perchlorate in fertilizers. However, it is possible to provide a snapshot of fertilizer commodities at any point in time.

This study was composed of two distinct phases. Phase 1 was designed to evaluate laboratory performance and the ruggedness of the method. Laboratory participation was on a voluntary basis. Laboratories were required to use ion chromatography, but were permitted to choose columns and operating conditions on their own, within certain limits. Phase 1 test samples included a wide variety of fertilizer matrices. In Phase 2, samples of materials were collected from around the nation and sent to the participating laboratories. All data were provided to EPA for evaluation and analysis. The final results and full details have been published in a report (*17*).

Phase 1: Evaluation of Participant Laboratories

A set of performance evaluation samples was prepared by the EPA. These spanned commodity chemicals, water-soluble plant foods, and granulated or pelletized lawn fertilizers. A combination of solid and liquid (aqueous) samples was sent to each laboratory. Laboratories were required to demonstrate recovery of fortifications, reproducibility, and ruggedness in real matrixes. They were also required to supply experimental details and calibration data. Lastly, quality control specifications were established with regard to number of replicates, blanks, spike recovery, accuracy, and precision. Laboratories were required to estimate limits of detection within each matrix using a standardized procedure (*18*).

The following materials were used as matrixes. Fortified samples and duplicates were included as well. Full details of composition are reported elsewhere (*17*). Among the matrix components were the following: granular triplesuperphosphate, kaolinite, bentonite, urea, ammonium nitrate, ammonium monohydrogen phosphate, potassium chloride, potassium sulfate, sodium chloride, potassium nitrate, tap water, deionized water, retail lawn fertilizers, water-soluble plant foods, and several mixtures of these.

Laboratories were required to demonstrate their ability to detect perchlorate in fertilizers. Some laboratories failed to adequately test aqueous samples with concentrations in the parts-per-billion, and were cautioned not to dilute samples initially. In Phase 1, the matrix was not identified to the laboratories. This made the process difficult, time-consuming, and resulted in higher errors because it was not possible to initiate corrective measures to handle the needs of a particular matrix. For this reason, major chemical constituents were identified for the laboratories in Phase 2.

The following laboratories successfully completed Phase 1: California Department of Food and Agriculture, Dionex Corporation, American Pacific Corporation, North Carolina State University Department of Soil Science, Montgomery Watson Laboratories, and IMC-Phosphates Environmental Laboratory.

Phase 2: Analysis of real world samples

Phase 2 consisted of the testing of fertilizer samples. The materials sampled represented current major suppliers of production farm or retail fertilizers and/or raw materials of consumer products. The following products were tested:

1. Multi-component fertilizers: lawn fertilizer (22-3-14), timed-release plant food (18-6-12), lawn fertilizer (36-6-6), water-soluble plant food (20-20-20), plantfood (10-10-10).
2. Macronutrient single-compound fertilizers: ammonium monohydrogen phosphate, urea, potassium chlorid e, ammonium sulfate, potassium sulfate, ammonium nitrate, sodium nitrate, potassium nitrate, potassium/sodium nitrate blends, granular triplesuperphosphate (hydrous calcium dihydrogen phosphates), ammonium dihydrogen phosphate.
3. Other additives and micronutrients: clay, potassium magnesium sulfate (langbeinite, mechanical mining and drill/blast), iron oxide.
4. Quality control samples

Positive hits above the preliminary assured reporting levels (pARLs) were rechecked. The pARL is a matrix-specific detection limit, which is described in the method (*18*). Perchlorate was detectable only in materials derived from Chilean caliche. Recoveries of fortifications ranged from 81% to 111%, regardless of the specific increase in concentration resulting from spiking. All laboratories demonstrated satisfactory performance in this regard. A full validation of the method and laboratory performance has been published (*19*).

It is worth pointing out at the perchlorate has been detected in isolated samples of sylvite taken from New Mexico (*20*). The only fertilizers unequivocally and consistently demonstrated to contain perchlorate were bagged products deriving nitrogen from Chilean nitrate salts, which are known to vary in perchlorate content. Perchlorate was undetectable in other fertilizer product tested by EPA. There is a consensus among researchers that there is insufficient evidence for fertilizers to be viewed as contributors to environmental perchlorate contamination, except for imported Chile saltpeter or products derived from it. The potential future influence of such products is further reduced by SQM's modified refining process to lower perchlorate concentration in its products.

Performance of the polyvinyl alcohol gel resin IC columns (100 mm and 150 mm) was evaluated (*21*). The NaOH eluent included an organic salt, sodium 4-cyanophenoxide. Detection was by suppressed conductivity. An archived set of Phase 2 samples was analyzed on the 150-mm column. The 100-mm column was used to further investigate the positive hits. Both columns gave satisfactory performance in fertilizer matrixes, with spike recoveries (±15%), assured reporting levels (0.5-225 $\mu g\ g^{-1}$ except for one at 1000 $\mu g\ g^{-1}$), accuracy (relative error <30% always and most <15%), and precision (injection-to-injection reproducibility <3% RSD) comparable to those reported in other studies. Performance did not vary substantially between column lengths. Lastly, the results of this investigation provided further evidence in support of the conclusions that had been reached previously by the EPA on the occurrence of perchlorate in fertilizers.

Implications for Vascular Plants

In the laboratory setting, some plant species will absorb perchlorate when exposed to perchlorate via irrigation water. This has been explored for possible phytoremediation (*22*). Some investigators have speculated that bacteria are responsible for perchlorate reduction in plants. Perchlorate-reducing monera have been identified by several laboratories, and cultured from a variety of sources (*23, 24*). This suggests that perchlorate-reducing bacteria are active in the environment.

Due to the reported occurrence of perchlorate in certain water resources and in certain fertilizer products, several groups have begun to address the extent and significance of perchlorate uptake by plants. For example, if produce is grown using perchlorate-tainted irrigation water or fertilizers and the perchlorate is retained in the edible portions, this might constitute a route of human exposure.

The possibility of exposure would be increased if perchlorate were shown to survive various types of processing. Unfortunately, experimental results that definitively gauge the extent of risk from this route of exposure have not yet been published. However, some progress toward this goal has been made.

One problem with uptake studies is the possibility of convolved influences on uptake. There are perchlorate absorption data available for only a few species of vascular plants. The absorption and accumulation of anionic solutes can be affected by many physical and chemical properties, such as concentration, size, charge density, and aquation. There are additional possible influence from soil sorption and or natural attenuation. These factors have been considered in some detail elsewhere (17).

To be applicable to agriculture or horticulture, studies of perchlorate absorption and accumulation must control for a variety of complicating factors. The presence of other anions either from co-administration in fertilizers or background salts present in the water supply may suppress uptake. Interspecies variation in absorption mechanisms may lead to differing levels of absorption and differing locations of accumulation. It is important to know if accumulation occurs in fruits versus in leaves or as a result of foliar application versus root application (as in irrigation).The rate of harvesting may lead to different rates of uptake by disrupting normal physiological processes in the plants. Lastly, the effect of soil (primarily sorptive in nature) must be considered. Careful agronomic studies are req uired to account for such influences, which are likely to complicate studies on the impact of contaminated irrigation water, too. Investigations are ongoing in these areas, but our understanding is currently incomplete and the kinds of generalizations that can be made are not especially useful for the construction of agricultural, horticultural, or environmental policy.

Initially, difficulty in analyzing plant tissues for perchlorate prevented some studies from being conducted. These problems have now more-or-less been overcome, and several reports detail how perchlorate content may be determined (13, 25, 26). Basic information on the analytical chemistry of perchlorate has been reviewed previously (14).

An obvious concern raised by finding measurable perchlorate concentrations in plant tissues is whether this ion can affect food crops. Most domestic crops are fertilized using commodity chemicals with no known link to perchlorate contamination. Some crops (e.g. corn, wheat, and rice) are fertilized with nitrogen fertilizers that should be perchlorate-free because of manufacturing processes. There is no reason to suspect perchlorate associated with growing grain.

The only crops with documented use of Chilean nitrate products are tobacco and citrus fruits. Data on application of perchlorate-containing fertilizers is sparse or nonexistent, and it is not possible to estimate the ecological impact in any meaningful way. Modest information is available on uptake by tobacco, but this is not a food crop, and the use of Chilean nitrate salts appears to be locale-dependent. The paucity of data makes it difficult to say whether imported produce is likely to be a source of perchlorate. Based solely on the figures for domestic use of Chilean nitrate salts, it hard to imagine that homegrown produce could ever be a major dietary source of perchlorate.

Even if many food plants can be shown to absorb and retain perchlorate under some conditions, the primary source of this contaminant is irrigation water, which is localized geographically. Accordingly, most of the country's agricultural products should be free from exposure via tainted irrigation water. On the other hand, some produce is largely supplied by regions that irrigate with Colorado River water, which is known to contain perchlorate. Therefore, such produce represents a potential exposure route for consumers. There are currently no investigations underway to examine food crops with documented exposure to perchlorate via irrigation or fertilization. While the likelihood of exposure via agricultural sources is small due to low consumption of Chilean nitrates and the low perchlorate concentrations therein, the significance of whatever exposure does occur is unknown in terms of food plant uptake or ecological impact.

References

1. Clark, J. J. J. In Perchlorate in the Environment; Urbansky, E.T., Ed. Kluwer Academic/Plenum: New York, NY, 2000; Ch. 3, and references therein.
2. Schilt, A. A. Perchloric Acid and Perchlorates. GFS Chemicals, Inc.: Columbus, OH, 1979, p 35, and references therein.
3. Ericksen, G. E. Am. Scien. 1983, 71, 366–374, and references therein.
4. Chilean Nitrate Corporation (CNC) website. 1999. URL: http://www.cncusa.com.
5. Searls, J. P. Potash-1999. U.S.G.S. Minerals Yearbook. 1999. URL:http://minerals.usgs.gov/minerals/pubs/commodity/potash560499.pdf
6. Urbansky, E. T.; Brown, S. K.; Magnuson, M. L.; Kelty, C. A. Environ. Pollut. 2001, 112, 299–302.
7. Environmental Protection Agency, Apache Powder Superfund Site: Status of Apache Cleanup Activities. St. David, Arizona. U.S. Environmental Protection Agency (Region 9): San Francisco, CA, May 1999.
8. Gullick, R. W.; LeChevallier, M. W.; Barhorst, T. S. J. – Am. Water Works Assoc. 2001, 93 (1), 66–77.

9. Lemons, J. F., Jr. Nitrogen. U.S.G.S. Minerals Information. 1996. URL: http://minerals.usgs.gov/minerals/pubs/commodity/nitrogen/ 480496.pdf.
10. Prud'homme, M. Potash. Natural Resources Canada. 1998. URL: http://www. nrcan.gc.ca/mms.cmy/content/1998/ 46.pdf.
11. Susarla, S.; Collette, T. W.; Garrison, A. W.; Wolfe, N. L.; McCutcheon, S. C. Environ. Sci. Technol. 1999, 33, 3469–3472.
12. Susarla, S.; Collette, T.W.; Garrison, A.W.; Wolfe, N.L.; McCutcheon, S.C. Environ. Sci. Technol. 2000, 34, 224.
13. Williams, T. L.; Martin, R. B .; Collette, T. W . Appl. Spectrosc. 2001, 55, 967–983.17
14. Urbansky, E. T. Crit. Rev. Anal. Chem. 2000, 30, 311–343.
15. Eldridge, J. E.; T sui, D. T.; Mattie, D. R.; Crown, J.; Scott, R. Perchlorate in Fertilizers. U.S. Air Force Research Laboratories: Wright Patterson AFB, OH, 2000; AFRL-HE-WP-TR-2000-0037.
16. Urbansky, E. T.; Magnuson, M. L.; Kelty, C. A.; Gu, B.; Brown, G. M. Environ. Sci. Technol. 2000, 34, 4452–4453.
17. Urbansky, E. T.; Collette, T. W.; Robarge, W . P.; Hall, W. L.; Skillen, J. M.; Kane, P. F. Survey of Fertilizers and Related Materials for Perchlorate (ClO4 –). Environmental Protection Agency: Cincinnati, OH, 2001; EPA/600/R-01/049.
18. Collette, T. W .; Robarge, W . P.; Urbansky, E. T. Ion Chromatographic Determination of Perchlorate: Analysis of Fertilizers and Related Materials. U.S. Environmental Protection Agency: Cincinnati, OH, 2001; EPA/600/R-01/026.
19. Urbansky, E. T.; Collette, T. W. J. Environ. Monit. 2001, 3, 454–462. 20. Harvey, G. J.; Tsui, D. T.; Eldridge, J. E.; Orris, G. J. 20th Annu. Meet. Abstr. Book – Soc. Environ. Toxicol. Chem., 1999; Paper no. PHA015, p. 277.
21. De Borba, B. M.; Urbansky, E. T. J. Environ. Monit. 2002, 4, in press.
22. Nzengung, V. A.; Wang, C. In Perchlorate in the Environment, E.T. Urbansky, Ed. Kluwer Academic/Plenum: New York, NY, 2000; ch 21. 23. Coates, J. D.; Michaelidou, U.; O'Connor, S. M.; Bruce, R. A.; Achenbach, L. A. In Perchlorate in the Environment; Urbansky, E.T., Ed. Kluwer/Plenum: New York, NY, 2000; ch 24, and references cited therein.
24. Logan, B. E. Environ. Sci. Technol. 2001, 35, 482A–487A, and references therein.
25. Ellington, J. J.; Evans, J. J. J. Chromatogr. A 2000, 898, 193–199.
26. Ellington, J. J.; W olfe, N. L.; Garrison, A. W.; Evans, J. J.; Avants, J. K.; Teng, Q. Environ. Sci. Technol. 2001, 35, 3213–3218.

Chapter 3

Perchlorate in Fertilizer? A Product Defense Story

Linda D. Weber[1], Wayne P. Robarge[2], William L. Hall, Jr.[1], and David Averitt[1]

[1]IMC Global Operations, 3095 County Road, Mulberry, FL 33860
[2]North Carolina State University, Box 7619, 3406 Williams Hall, Raleigh, NC 27695-7619

In 1999, a paper entitled Perchlorate Identification In Fertilizers was published which implicated a number of lawn and garden fertilizers and fertilizer source materials as containing perchlorate (1). After re-analyzing these samples using improved sampling and analytical protocols, most of the fertilizer source materials were found not to contain perchlorate (non-detectable amounts). Perchlorate was still present, to a lesser extent, in several fertilizer source materials and all of the original lawn and garden products. However, no perchlorate has been found in any subsequent analyses of the same products that were later purchased (2). To further investigate this issue, IMC-Global, the world's largest producer and marketer of concentrated phosphate and potash crop nutrients for the agricultural industry, initiated its own analysis program to survey for the possible presence of perchlorate in fertilizer source materials and lawn and garden products, and in samples archived as part of its quality control program. Additional historical samples were recovered from the Magruder check sample program managed through AAPFCO (3). Lastly, IMC Global also participated in several round-robin studies organized by the US Environmental Protection Agency and The Fertilizer Institute (4) to systematically evaluate whether perchlorate does or should be expected to occur in fertilizers.

Introduction:

Perchlorate contamination in water supplies has recently become a significant public health issue (5). Perchlorate is readily absorbed into the gastrointestinal tract after ingestion. The perchlorate ion is similar in size to iodide, and competes with iodide for uptake by the thyroid gland. A reduction in iodide uptake by the thyroid results in low thyroid hormone production and diseases associated with this condition. Perchlorate is currently on the Unregulated Contaminant Monitoring Rule (UCMR) List 1 for assessment monitoring. California's provisional action level is 18 µg L^{-1} in drinking water. Of these, 38 wells have perchlorate concentrations above the action level. California water suppliers have detected perchlorate in 144 public water supply wells. There are 19 states with confirmed releases of perchlorate in ground and surface water.

Perchlorate is an anion available as a salt with various cations. The most common forms include ammonium perchlorate (used as a solid rocket oxidant and ignition source in munitions and fireworks), potassium perchlorate (used in air bag inflation systems, road flares, and as a medication to treat hypothyroidism), and perchloric acid. In the environment, perchlorate is very mobile and persistent in surface and ground waters (6). Perchlorate has been released into the environment primarily from manufacturing facilities and solid rocket booster testing and maintenance sites.

The only known natural occurrence of perchlorate has been associated with Chilean nitrate deposits. Nitrate fields in Chile occur in areas of low relief with rounded hills and rides and broad shallow valleys. Deposits occur as veins and impregnations in host rocks and in unconsolidated sediments. Annual precipitation in this region is 50 mm or less (7).

Ion chromatography has emerged as the method of choice for analysis of perchlorate in drinking water and was applied to fertilizer extracts (8). EPA 314.0 is currently the approved ion chromatographic procedure for analysis of perchlorate in drinking water. Recently an EPA procedure for analysis of perchlorate in fertilizers using ion chromatography has been released (9). The need for this procedure was evident as early reports of perchlorate in fertilizer were essentially based on EPA 314.0, which is not appropriate for the complex matrix represented by fertilizers. The inconsistencies in various data reported for the same fertilizer products illustrated the need for standardized sampling, extraction and analyses when handling this material.

Background

In 1999, a paper entitled Perchlorate Identification In Fertilizers (1) was published which implicated a number of Lawn and Garden fertilizers and fertilizer source materials as containing perchlorate, with the reported concentrations ranging from 0.1% to 3.64%, (Tables I and II). It has been known for some time that sodium nitrate rich caliché ores in Chile are a naturally occurring source of perchlorate (9, 10). However, the presence of perchlorate in other sources especially phosphate rock and urea, was unexpected. The source of the Florida phosphate rock and western rock listed in Table 1 was the National Institute of Standards and Technology (NIST), whose certified analysis was inconsistent with the reported perchlorate concentrations. The NIST was also the source of the dihydrogen ammonium phosphate sample, which was actually made by J. T. Baker Corporation (11). The reported presence of perchlorate (at 0.46%) in this pure analytical grade material was also inconsistent with the certified analysis (99.96% dihydrogen ammonium phosphate based on P content). Lastly, urea is formulated by a reaction between gaseous ammonia and carbon dioxide, leaving essentially no chance that perchlorate could be a contaminant in this material at levels reported in Table I.

Following discussions with the authors (1), representatives from academia, other USEPA laboratories, and the fertilizer industry, an erratum (12) was published November 1, 2000, stating that six values in Table I of the article were incorrectly reported. Those values were based only on ion chromatography using a Dionex^{TM} AS-11 column. Corrected values (Table III) were obtained using a Dionex^{TM} AS-16 column designed for perchlorate analysis (13), and confirmed with spikes and also by capillary electrophoresis.

TABLE I. Perchlorate Concentrations in Fertilizer Components

component	perchlorate (%)
phosphate rock (western) [a]	0.10 ± 0.01 [c]
phosphate rock (Florida) [a]	0.11 ± 0.02
potash (commercial) [b]	0.29 ± 0.03
potash (muriate) [b]	0.36 ± 0.04
dihydrogen ammonium phosphate [a]	0.46 ± 0.05
urea [d]	0.25 ± 0.02
langbenite [e]	1.86 ± 0.21
Chilean nitrate	3.64 ± 0.34

[a] Samples were purchased from National Institute of Standards and Technology (NIST), Gaithersburg, MD. [b] Commercial source of potash samples. Provided by Greg Harvey, Wright-Patterson AFB, from the mineral archives of the U.S. Geological Survey. [c] plus or minus sign (±) indicates the deviation in measurement among the six replicates.
[d] Urea sample was purchased from Goldkist, Commerce, GA. [e] TRC, Irvine, CA.

TABLE II. Perchlorate Concentrations in Commercial Fertilizers[a]

brand/description (N-P-K)	manufacturer/ lot no.	perchlorate (%)
ammonium nitrate 34-0-0	Gold Kist Commerce, GA 20631	0.22 ± 0.04 [b]
Lesco 17-3-11	Lesco Rocky River, OH 023-371	0.57 ± 0.03
Procare 10-10-10	Gro Tech Inc. Madison, GA 525-1123	0.20 ± 0.08
fallfeed winterizer 18-6-12	Purcell Industries Sylacauga, AL F 1061-1169	0.15 ± 0.08
STA-Green 12-6-6	Purcell Industries Sylacauga, AL 1061-1324	0.84 ± 0.18
Scotts winterizer 22-4-14	The Scotts Co. Marysville, OH FL 6432037	0.51 ± 0.05
Vigaro 10-10-10	Gro Tec Inc. Madison, GA 525-1123	0.55 ± 0.06
premium lawn 27-2-5	Vigaro Industries Winter Haven, FL 735-8512	0.33 ± 0.08
Pennington 34-0-0	Gro Tech Inc. Madison, GA 525-2043	0.61 ± 0.04

[a] Samples were analyzed by ion chromatography with six independent replicates. Perchlorate given in wt %. [b] plus or minus sign (±) indicates the deviation in measurement among the six replicates.

Table III. Corrected Perchlorate Concentrations in Fertilizer Components

component	perchlorate (%)
phosphate rock (Western)	ND[a]
phosphate rock (Florida)	ND
potash (commercial)	0.004[b]
potash (muriate)	ND
dihydrogen ammonium phosphate	ND
urea	ND

[a] ND, not detected by IC above 0.003%. [b] Potash (commercial) was confirmed by nuclear magnetic resonance (NMR) spectroscopy.

Reanalysis of the remaining samples listed in Table I confirmed the presence of perchlorate in the liquid extracts from langbeinite ore and Chilean nitrate but at much reduced levels. In addition, presence of perchlorate was reconfirmed in the lawn and garden samples originally reported in Table II. The existence of another study commissioned by the Perchlorate Working Group (8) also became known in more detail. In this study 30+ liquid extracts, primarily of lawn and garden fertilizers, were reported to contain perchlorate at levels approaching 0.5%. These samples were reportedly purchased during the same approximate times as the samples listed in Table II.

The reported existence of perchlorate in lawn and garden fertilizers represented in Table II, and reported in the study commissioned by the Perchlorate Working Group, proved even more perplexing as efforts to verify these observations by other laboratories failed (9, 12). These attempts were on fertilizers purchased following publication of the initial results. All attempts to verify or reproduce the initial findings by purchasing new samples have essentially failed with the exception of products containing Chilean nitrate. Indeed, the same authors of the original report later purchased 17 additional fertilizer mixtures, many the same products in Table II, during the period of August – September 1999, and detected perchlorate in only one sample using ion chromatography (2). This observation could not be confirmed using Raman spectroscopy. These later observations are consistent with other reports, on a growing number of analyses, on a variety of fertilizer materials, failing to detect the presence of perchlorate, except in materials which were known to be derived from Chilean caliché (12).

Because of the difficulty resolving the issue of perchlorate in fertilizers, IMC Global, the world's largest producer of phosphate and potash nutrients, initiated its own analytical program to monitor the possible presence of perchlorate in fertilizers. In particular, efforts were directed at identifying and isolating archived samples through internal quality programs, or using Magruder check sample program (3) samples to determine how wide spread the supposed occurrence of perchlorate in fertilizers was in 1999. Was this reported presence of perchlorate a truly unique one-time phenomenon unlikely to be repeated, or was it the result of some other as of yet unknown set of factors unrelated to production of fertilizers? Additionaly, it was deemed important for IMC-Global to participate in subsequent efforts to improve the methodology for determining perchlorate in fertilizer matrices.

Analytical Protocols

Equipment

A Dionex model DX-600 Ion Chromatograph is used in the analysis of perchlorate in fertilizer samples. The system includes a GS-50 gradient pump (operated in isocratic mode), LC-30 chromatography oven, a CD-25 conductivity detector, an EG-40 KOH eluent generator, and an AS-40 autosampler driven by Peaknet 6.0 software.

Background suppression is achieved using an anion self-regenerating suppressor (ASRS-II; (4mm)) with a current setting at 300 mA, operated in external mode. The sample loop is set at 1000 uL, and 50 mM potassium hydroxide is produced by the eluent generator using 18 megaohm water degassed with helium for approximately 15 minutes prior to use. Eluent flow rate is set to 1.5 mL/min. The concentration of perchlorate is calculated based on peak area.

The AS-16 analytical column and AG-16 guard column is used for all analyses, although previous studies used the AS-11 analytical column. The AS-16 was designed specifically for this analysis, and is currently the column of choice due to superior peak resolution and reduced matrix effects (14).

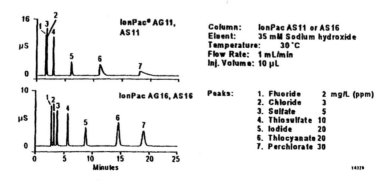

Figure 1. Comparison of the AS-11 and AS-16 Columns (13)
{Courtesy of Dionex Corporation}

Analytical Method
Standards are prepared from reagent grade sodium perchlorate. All solutions are prepared from 18 megaohm water. Samples are prepared referencing the procedure recommended in the 2000 round robin study (9) for the determination of perchlorate in fertilizers. The procedure calls for 1:10 fertilizer mass-to-leachate/solution volume ratio using pulverized solid material (except nitrate salts), followed by 8-15 hours of vigorous mixing. Visible suspended matter in the liquid portion is removed by filtration or centrifugation. Filtrates should be initially run with a 1/1000 (0.1%) v/v dilution. If no peak is visible, the 10% dilution, 1% dilution, or the original solution may be run at the discretion of the analyst (no undiluted original solutions were injected for this study due to the detrimental effects on instrument consumables). All samples are spiked with a known amount of perchlorate, and must have 80-120% recovery to be acceptable, otherwise further dilution was necessary.

Method Detection Limit and Calibration
To determine the instrument MDL, eight replicates of a 3 ug/L perchlorate standard were analyzed. Student's t for eight samples (n=8, v=7 degrees of freedom) at the 99% confidence limit was 2.998. At this detection limit, the peak

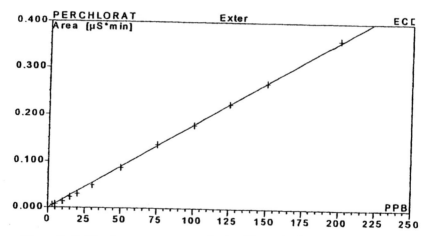

Figure 2. Calibration curves generated using 12 standards and AS-16 analytical column. (Correlation coefficient = 0.9998)

was resolved and signal to noise ratio was greater than 3. An instrument detection limit of 1.44 ug/L was calculated based on peak areas with an average retention time of 9.10 +/- 0.02 minutes. The solid phase MDL was done by back calculating from the aqueous solution MDL, taking into account a 1:10 sample mass to volume ratio, and was determined to be 14.4 ng ClO_4^-/g fertilizer. A typical standard curve derived using 12 standards is illustrated in Figure 2.

Results and Discussion

Analytical Capability

IMC Global participated in a series of sample exchanges arranged by the US Environmental Protection Agency and The Fertilizer Institute (4, 8). The express purpose of this sample exchange was to confirm the reliability of using ion chromatography to identify the presence of perchlorate in a complicated sample matrix such as generated by fertilizer source materials and fertilizer blends. In all cases (30+ different sample matrices and spiked samples with perchlorate), the results obtained by IMC-Global were comparable to those of the other participating laboratories in terms of accuracy, spike recovery and detection limit. (See (8) for detailed comparison between participating laboratories and anlysis of results.) The IMC-Global laboratory was deemed competent, therefore, to detect the presence of perchlorate in fertilizer matrices using ion chromatograph and the analytical protocol described above.

Magruder Check Samples

A total of 30 individual Magruder Check Samples were located and identified spanning the time period of 1993 to 1999. The majority of samples were prepared in 1997, 1998, and 1999 (Table IV; the first two numbers identify the production year, the third and fourth numbers indicate the production month). For the samples listed, one or more were prepared monthly by an independent laboratory from typical materials supplied by numerous manufacturers. These materials are the same as those available in the retail market. The names and affiliations of the individuals who actually prepared the samples are indicated in the column marked History (Table IV).

Of the 30 samples located and analyzed, only one (9711 B) was positive for the presence of perchlorate. This sample was composed entirely of potassium nitrate which is derived from Chilean caliché. Chilean caliché, as already noted, is a known source of naturally occurring perchlorate. No perchlorate was found in the lawn and garden formulations tested from 1998 (n=3) and 1999 (n=7), which taken together, exceeds the number of materials tested in 1999 and reported to contain perchlorate (1,2).

Table IV. Perchlorate Content of Magruder Check Samples (1993 - 1999)

Magruder No.[a]	N-P-K Analysis	Perchlorate mg/kg	History
9311-B	0-0-61 Pot. Chloride	ND[b]	Bill Hall, Vigoro Industries
9406-B	0-0-22 (Langbeinite)	ND	Jim Smith
9509	2-5-20 + minor elem.	ND	Bill Hall, Vigoro Industries
9804	10-10-10 All Purpose	ND	Sandy Simon, Pursell Ind.
9805	20-5-15 Vig. Lawn Fert.	ND	Bill Hall, IMC Vigoro
9806	6-24-24 Ag Granulated	ND	Pat Peterson, CF Industries
9812	5-10-15 Garden Fertilizer	ND	Don Day, PCS Sales
9901	18-4-10 Lawn Fertilizer	ND	Greg Haberkost, Lebanon
9902	20-20-20 All Purpose W/S	ND	Bill Hall, IMC Vigoro
9905	20-4-4 Lawn Fertilizer	ND	Greg Haberkost, Lebanon
9907-A	Phosphate Rock	ND	CF Ind, Harold Falls
9907-B	40-0-0 Methylene Urea	ND	Dick Harrell, NuGro
9908	19-19-19 All Purpose	ND	Bob Beine, U of Ky
9909	27-3-5 Scotts Turfbuilder	ND	V. Snyder, Scotts Co.
9911	10-20-20 Garden Fert.	ND	Greg Haberkost, Lebanon
9912	13-13-13 Gran. All Purp.	ND	B. Avant, Royster Clark
9607	29-3-6	ND	Bill Hall, IMC-Vigoro
9608	16-4-8	ND	Jim Smith
9609	20-0-20	ND	Bill Hall, IMC-Vigoro
9611	15-15-15	ND	Bill Hall, IMC-Vigoro
9612	8-16-24	ND	Jim Smith
9701B	0-0-22 (Langbeinite)	ND	Mabry Handley
9702	16-16-16	ND	Bill Hall, IMC-Vigoro
9707	16-04-08	ND	Greg Haberkost (Lebanon)
9708	29-03-04	ND	Sandy Simon (Purcell)
9709	13-13-13	ND	Bobby Avant (IMC)
9710B	00-00-62 KCl	ND	Bill Hall, IMC-Vigoro
9711 B	13-00-44 Potassium Nitrate	1900	Bill Hall, IMC-Vigoro
9803A	0-0-22 (Langbeinite)	ND	Sandy Simon, Purcell
9810B	00-00-50 Potassium Sulfate	ND	Bill Hall, IMC-Vigoro

[a] *The first two numbers identify the production year, the third and fourth numbers indicate the production month. "A" and "B" designations indentify multiple samples within the same month.* [b] *ND = not detected.*

Fertilizer Source Materials Produced by IMC Global

IMC-Global is a global producer of potash and fertilizer source materials derived from langbeninite ore. Both of these materials were reported as containing perchlorate (Table I). A total of 29 internal quality control samples were obtained and analyzed from IMC-Global production facilities located in Carlsbad, NM, and at three production facilities in Canada (Table V). These samples span the period 2000 to 2002. No perchlorate was detected in these samples by the IMC-Global laboratory using ion chromatography control. A subset of these samples sent to another laboratory for analysis (Dr. Wayne P. Robarge, North Carolina State University, Raleigh, NC) also confirmed the absence of perhclorate in these materials as determined using ion chromatography.

Table V. Potash and Langbeinite Ore Samples Tested for Perchlorate.[a]

Carlsbad NM:
Langbeinite Ore, IMC Carlsbad area 7, Drill & Blast Mining method (Aug. 2000)
Langbeinite Ore, IMC Carlsbad area 9, continuous mining method, (Aug. 2000)
Carlsbad KMAG Process Dryer Feed, process dispatch bin feed, (August 2000)
Langbeinite ore sample-mechanical mining method, (2000)
Langbeinite finished product, (2000, 2001, 2002)
Granular K-mag (2001, 2002)
Granular SOP (2001, 2002)
Premium K-mag (2001, 2002)
Sylvite (2001, 2002)
Langbeinite ore (2001, 2002)

Esterhazy, SK:	**Colonsay, SK:**
Ore (2001, 2002)	Potash Ore, 2000, 2001, 2002
Granular (2001, 2002)	**Belle Plaine, SK:**
Course (2001, 2002)	Granular KCl, 2000, 2001 2002

[a] Perchlorate was absent in all samples as tested by ion chromagraphy.

Commerical Lawn and Garden Fertilizers

A total of 7 bags of lawn and garden fertilizers were purchased in the Valrico, Florida area at commercial outlets in November 2001 (Table VI). These fertilizers were similar in composition to those reported to contain perchlorate in 1999 (1,2). As observed for the Magruder Check Samples, only the product derived from Chilean caliché (Nitrate of Soda) tested positive for perchlorate.

Table VI. Perchlorate Content of Lawn and Garden Fertilizers (Nov. 2001)

N-P-K Analysis	Perchlorate mg/kg	Product ID
14-14-14	ND	Osmocote
20-27-5	ND	Scotts Veg. Food
20-20-20	ND	Peters All Purpose
12-10-5	ND	Vigoro
15-30-15	ND	Miracle Gro
16-0-0	1200	Nitrate of Soda
30-3-3	ND	Scotts Turf Bldr.

ND = not detected.

Conclusion

IMC-Global is the world's largest producer and marketer of concentrated phosphate and potash crop nutrients. The possible presence of perchlorate in these products and the potential implications for human health required immediate action by IMC-Global to address this issue and determine the cause of the reported observations of perchlorate in fertilizers, especially commercial lawn and garden fertilizers. In-house analytical capability for perchlorate analysis in fertilizer analyses has been established and shown through extensive inter-laboratory studies to be fully functional and comparable to other recognized laboratories in its ability to detect the presence and/or absence of perchlorate in fertilizer materials. Application of this analytical capability for analysis of historical samples of fertilizer materials, and other current samples from ore bodies and derived fertilizer products has failed to detect the presence of perchlorate in any material, except those known to contain or are derived from Chilean caliché. The presence of perchlorate in the wide variety of fertilizer materials as initially reported has been described as a "…phenomenon (that) appears to have constituted a sporadic-if not singular event rather than reflecting a recurrent problem." (14) Our data do not support such a conclusion given the reported occurrence of perchlorate that seemingly appeared in every product purchased, regardless of manufacturer represented in the initial sample collection. Nor does this conclusion seem logical in light of the fact that Chilean caliché makes up less than 0.1% of the fertilizer tonnage marketed in the United States. It remains a mystery that our historical samples also failed to reflect this seemingly pervasive presence of perchlorate in the lawn and garden products purchased across the United States in 1998-1999. This paper cannot provide an explanation as to why perchlorate was detected in a select group of fertilizer materials. It does, however, contribute to the growing database that supports the conclusion that fertilizer source materials were not the source of perchlorate in the original products.

Acknowledgements:

The assistance of Richard Dorman, Coastal Chemical Inc. and Dr. Eleanour Snow, University of South Florida in gathering reference materials for this report is recognized. Dr. P. E. Jackson, Dionex Corporation, David E. Gadsby IMC-Phosphates and Dennis G. Sebastian IMC-Phosphates, for technical assistance. Annette Revet and Steve Gamble, IMC-Kalium, Harold Falls, CF Industries, Dr. Lauterbach, SQM, W. Hall, IMC-Global, for supplying samples.

References

1. Sursala, S., Collette, T. W., Garrison, A. W., Wolfe, N.L. and McCutcheon, S. C. Perchlorate in Fertilizers, Environmental Science and Technology, **1999**, 33, 3469-3472.

2. Sursala, S., Collette, T. W., Garrison, A. W., Wolfe, N.L. and McCutcheon, S. C. Perchlorate in Fertilizers, Additions and Corrections- Environmental Science and Technology, **2000**, 34.

3. AAPFCO Association of American Plant Food Control Officials Official Publication No. 56, **2003**; Joel Padmore, Treasurer NC Department of Agriculture 4000 Reedy Creek Rd Raleigh NC 27607-6468

4. The Fertilizer Institute, Union Center Plaza 820 First Street, N.E. Suite 430 Washington, D.C. 20002

5. Clark, J.J.J. Toxicology of Perchlorate. In Perchlorate in the Environment, E.T. Urbansky, Ed. Kluwer/Plenum: New York, NY, **2000**; chapter 3, and references cited therein.

6. US EPA, Perchlorate environmental contamination: toxilogical review and risk characterization based on emerging information. US EPA Office of Research and Development, National Center for Environmental Assessment, Washington DC, **2002**

7. Orris, G. J., Bliss, J.D., Some Industrial Mineral Deposit Models: Descriptive Deposit Models; U.S. Department of the Interior, U.S. Geological Survey, Open File report 91-11A, Tucson, AZ, **1991**.

8. Eldridge, J. E.; Tsui, D. T., Mattie, D. R., Crown, J.; Scott, R. <u>Perchlorate In Fertilizers.</u> U.S. Air Force Research Laboratories: Wright Patterson AFB, OH, **2000**. Report in press.

9. Urbansky, Edward T., Collette T. W., Robarge, W. P., Hall W. L, Skillen J. M., Kane, P. F. <u>Survey of Fertilizers and Related Materials for Perchlorate (ClO_4^-),</u> U. S. Environmental Protection Agency, National Risk Management Research Laboratory, Office of Research and Development. Cincinnati OH, May **2001**.

10. A. A. Schilt, <u>Perchloric Acid and Perchlorates</u>, GFS Chemicals, Inc., Columbus, OH, **1979**

11. National Institute of Standards and Technology, <u>Certificate of Analysis Standard Reference Material 194</u>, September 1, **1992**

12. Urbansky, Edward T. <u>Comment on "Perchlorate Identification in Fertilizers" and the Subsequent Addition/Correction.</u> , Additions and Corrections- Environmental Science and Technology, **2000**, 35.

13. P. E. Jackson S. Gokhale and J. S. Rohrer, in <u>Perchlorate in the Environment</u>, ed. E. T. Urbansky, Kluwer/Plenum, New York, **2000**, ch 5.

14. P. E. Jackson, S. Gokhale, T. Streib, J. S. Rohrer and C. A. Pohl, <u>J. Chromatogr., A.</u> **2000**, 888, 151

Chapter 4

Reduction of Perchlorate Levels of Sodium and Potassium Nitrates Derived from Natural Caliche Ore

A. Lauterbach

SQM, Anibal Pinto 3228, Antofagasta, Chile

SQM is a manufacturer of specialty nitrate fertilizer products used in the US for very specific and specialized purposes. Sales of SQM products make up about 0.1% of the total US fertilizer market. Some of these products contain low levels of naturally occurring potassium perchlorate found in the caliche ore from which they are derived. Upon learning of the potential health effects of perchlorate containing compounds in the environment and in response to customer inquiries concerning perchlorate, SQM conducted successful studies to reduce perchlorate in its products. Data indicates that most contamination results primarily from defense and industrial operations in the western US, consequently SQM has no reason to believe fertilizer use has affected in any way public health or the environment. Regardless, SQM has already begun producing and selling fertilizers with perchlorate levels below 0.01% w/w. The traditional production processes for sodium and potassium nitrate, and the process modifications developed for reducing perchlorate content will be presented.

BACKGROUND AND BRIEF HISTORY

Commercial use of the caliche ore deposits in northern Chile began in the 1830's, when sodium nitrate was extracted from the ore for use in the manufacture of explosives and fertilizers. By the end of the nineteenth century, nitrate production had become the leading industry in Chile and the country was the world's leading supplier of nitrates[1]. The accelerated commercial development of synthetic nitrates in the 1920s and the global economic depression in the 1930s caused a serious contraction of the Chilean nitrate business, which did not recover significantly until shortly before the Second World War. After the war, the widespread commercial production of synthetic nitrates resulted in a further contraction of the natural nitrate industry in Chile, which continued to operate at depressed levels into the 1960s.

SQM was established in 1968, and acquired its then principal properties from Anglo Lautaro and Corfo, the Chilean state-owned development corporation. In 1971, Anglo Lautaro sold all of its shares of SQM to Corfo and SQM remained wholly owned by the Chilean government until 1983. In 1983, Corfo began the privatization of SQM with the sale of SQM's shares to the public. In subsequent years, Corfo sold additional shares of SQM and, by 1988, all of SQM's shares were owned by the private sector. In September 1993, SQM established its American Depositary Receipt (ADR) program and its shares were listed on the New York Stock Exchange as ADR's.

The Company

SQM is an integrated producer of specialty fertilizers, iodine, lithium carbonate, and a producer of certain industrial chemicals, including industrial nitrates. SQM's products are derived from the unique mineral deposits found in the Atacama desert region of northern Chile, where the Company mines and processes caliche ore and brine deposits. The caliche ore contains the largest known nitrate and iodine deposits in the world. The brine deposits of the Atacama Salar contain the highest known concentrations of lithium and potassium as well as significant concentrations of sulfate and boron.

From its caliche ore deposits, SQM produces a wide range of nitrate-based products, used for specialty fertilizers and industrial applications, as well as iodine and iodine derivatives. At the Salar de Atacama, SQM extracts the brines rich in potassium, lithium and boron and produces potassium chloride, potassium sulfate, lithium solutions, and boric acid. SQM produces lithium carbonate at a plant near the city of Antofagasta, Chile, from the solutions brought from the Salar de Atacama.

Use of SQM Nitrates in the United States

SQM's nitrate fertilizers make up <0.1% of the US fertilizer market(2). Some SQM products contain varying levels of naturally occurring perchlorate, resulting from the caliche ore from which they are derived. Studies have detected perchlorate in SQM product at concentrations of about 2 mg/g (3).

Upon learning of potential health effects of perchlorate (resulting primarily from defense and industrial operations in the western US), and in response to customer inquiries, SQM conducted successful studies to develop methods to reduce perchlorate in its products. Although SQM has no reason to believe fertilizer use has affected in any way public health or the environment, the company now produces fertilizers with perchlorate levels below 0.01% w/w.

The nitrate deposits (see Figure 1) are located in northern Chile, in a plateau between the costal range and the Andes mountains, in the Atacama desert. They are scattered in an area extending some 700 km long, and ranging in width from a few km to 50 km. Most deposits are in areas of low relief, about 1,200 m above sea level. The nitrate ore is a conglomerate of insoluble and barren material (breccia, sands and clays) firmly cemented by soluble oxidized salts.

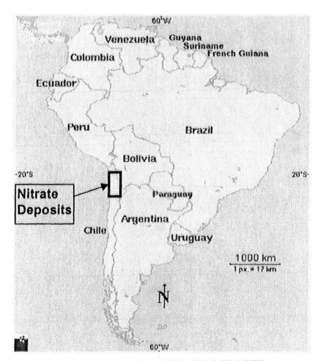

Figure 1. CALICHE: NITRATE DEPOSITS

The deposits are composed of several layers, and are very heterogeneous, being variable in size, thickness, composition and hardness. Overburden may include "chuca", a layer of unconsolidated sand, silt and clay, and "panqueque", semi consolidated porous material poorly cemented by salts, over loose gravel. The ore composition has degraded considerably since the 1830's, when it was reported that ores of 50% sodium nitrate were mined. Presently there are still reserves that can be mined for several more decades. The average content of saline water soluble constituents is shown in Table 1 (SQM Internal Data).

Table 1. Analysis of Nitrate Ore Samples

Component wt%	Mining Period		
	Up to 1920	1960 - 1970	1970 - 2000
$NaNO_3$	20 -50	7 - 9	6.5 - 8
NaCl	20	10 - 15	5 - 10
Na_2SO_4	12 - 15	12 - 16	12 - 20
I_2	0.03	0.03	0.03
$Na_2B_4O_7$	0.4- 0.6	0.4 - 0.6	0.4 - 0.6
K	1.0 - 2.0	0.4 - 1.5	0.3 - 1.2
$KClO_4$	0.03	0.03	0.03
Mg	0.2 - 0.3	0.2 - 0.8	0.2 - 1.2
Ca	0.4 - 3.3	0.4 - 2.3	0.5 - 2.8
H_2O	1.0 - 2.0	1.1 - 1.8	1.1 - 1.8
Insolubles	6 - 14	53 - 68	60 - 69

Origin of the Deposits

The Atacama desert of northern Chile, where the nitrate-rich caliche deposits were formed, is the driest of the world's deserts. Average annual rainfall is normally less than 1 mm/year. Measurable rainfall (1 mm or more) may be as infrequent as once every 5 to 20 years. Many theories about the origin of the Chilean Nitrates deposits have been proposed. Charles Darwin, visited the deposits in northern Tarapacá in July 1835 during the voyage of the *Beagle*. He speculated that they were formed at the margin of an inland extension of the sea (4). Almost every conceivable source and mode of accumulation for the nitrate has been suggested; most theories either ignore the sources of other constituents or assume they are compatible with the proposed source of nitrate.

According to Ericksen*(5)* nitrate deposits likely formed during the Tertiary and Quaternary periods (last 10-15 million years) by accumulation of saline components by deposition from the atmosphere of materials from diverse local rather than distant sources. Among these, the ocean was likely important. Emanations of rocks from volcanism in the northern Andes Mountains were also important sources of some saline components. Photochemical reactions in the atmosphere probably played a role in the formation of nitrate, sulfate, iodate, and perchlorate, all constituents of the deposits.

PRODUCTION OF AGRICULTURAL GRADE SODIUM NITRATE AND POTASSIUM NITRATE

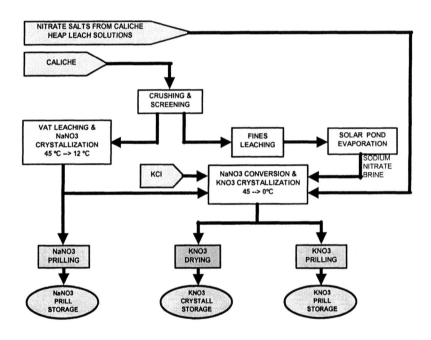

Figure 2. Agricultural Grade Sodium and Potassium Nitrate Production Process

Production of agricultural nitrate fertilizers begins by removal of fines from crushed ore (Figure 2). The material is leached in a series of 10,000 m^3 leaching vats, with the mother liquor having between 300 and 350 g/l of sodium nitrate and other soluble components of the ore. After circulating sequentially through the vats, enriched nitrate solution, containing about 450 g/L sodium nitrate, is cooled to 8 – 12 °C crystallizing essentially pure sodium nitrate.

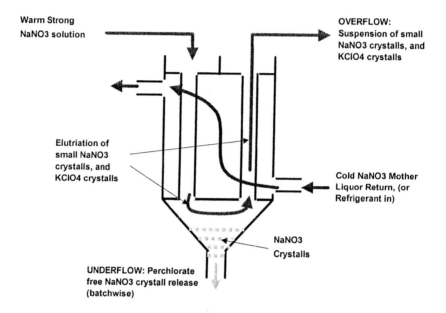

Figure 3. Characteristics of the sodium nitrate crystallizers

Analyzing the characteristics of the sodium nitrate crystallizers, it was determined that the underflow sodium nitrate pulp (see Fig. 3) contained practically no potassium perchlorate crystals. This is due to the relatively high rate of flow of the warm sodium nitrate solution through the heat exchanger tubes. As a consequence of this high rate of flow, the potassium perchlorate crystals, having a smaller particle size than the mean size of the sodium nitrate crystals, are elutriated and therefore removed together with the fraction of small sodium nitrate crystals. Some changes in the present plant layout will permit the separation of these two fractions, increasing substancially the proportion of low perchlorate sodium nitrate by increasing wash water into the solid / liquid separation step of the underflow fraction.

Further examination of the existing crystallizers suggested the perchlorate content of the crystallized sodium nitrate could be reduced if the crystallization temperature was increased 12 hours per day. By using this modification a small amount of low perchlorate crystalline sodium nitrate was produced. However, this change also reduced yields of sodium nitrate. Increasing the crystallization temperature for more than 12 hours per day results in the establishment of a new equilibrium and perchlorate precipitation with crystallized sodium nitrate. Only by increasing the sodium nitrate crystallization temperature, can perchlorate crystallization be avoided during continuous operation. However, this change reduces the nitrate yield substantially, rendering the process non-profitable.

PRODUCTION OF TECHNICAL AND REFINED GRADE POTASSIUM NITRATE

Technical and refined potassium nitrate is produced by reacting sodium nitrate and potassium chloride at 80°C (conversion reaction), precipitating sodium chloride that is discarded (see Figure 4). The potassium nitrate solution is crystallized at 40 °C, yielding technical potassium nitrate. By submitting the crystals to a refining/washing step, refined potassium nitrate is obtained.

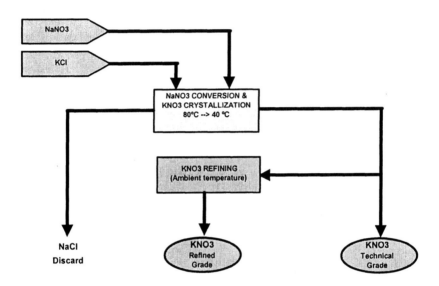

Figure 4. Technical and Refined Grade Potassium Nitrate Production Process

The process in Figure 4 was designed to obtain quantities of pure potassium nitrate, chloride is the main impurity. Technical grade has a Cl content of <0.2%, the refined grade <0.03%. Keeping chloride concentration of the mother liquor below a target concentration, affects the efficiency of the washing step during centrifugation (solid/liquid separation). Wash water and corresponding purge are adjusted to maintain chloride specifications for each grade.

It has been determined that perchlorate content of the product is dependent on the proportion of ClO_4^- to Cl^- in the mother liquor. The technical product guarantees a maximum Cl^- content of 0.2%, yielding a maximum ClO_4^- content of 0.01%. Therefore, if the ClO_4^-/Cl^- ratio is larger than or equal to 0.01/0.2 in the wash water, the corresponding purge to the solar ponds has to be adjusted to minimize inclusion of perchlorate in the product.

FORMULATING LOW PERCHLORATE CONTAINING PRODUCTS

Potential health effects due to the presence of perchlorate, result primarily from defense and industrial sources *(7)*. Although SQM does not believe fertilizer use negatively affects human health or the environment, it has developed methods to limit perchlorate in its products. Attempts to alter current process streams by changing crystallization temperatures permitted production of small amounts of low perchlorate product. However, this process needs additional washing to be effective. Although not difficult to implement, it proved to be economically unsound. An alternative is separation of the sodium nitrate fractions (in Figure 3). It was necessary to conduct a full technical and economic evaluation to develop a method reducing perchlorate content of SQM products below 0.01%, a level equivalent to technical grade potassium nitrate.

For agricultural-grade potassium nitrate, the most successful approach to limit perchlorate is consistent with controlling the maximum perchlorate content of the mother liquor during production of technical product. That is, as long as the production process keeps the ratio between ClO_4^- and Cl^- equal or less than 0.05, the perchlorate content will be les than 0.01% and the chloride content of the product will remain below the guaranteed upper limit (<0.2% as chloride).

PROCESS SCHEME TO INCREASE PRODUCTION

The two production schemes outlined succeed in producing low-perchlorate content products. However, plant efficiencies reduced.

Two alternative schemes are currently under review to reduce perchlorate content of SQM's total nitrate production. The first alternative is based on the specific crystallization processes (see Figure 5). It consists of separating high and low perchlorate fractions in the present sodium nitrate crystallization process (see Figure 3 and Figure6), refining the high perchlorate sodium nitrate fraction at high temperature (75 °C), and re-crystallizing the agricultural grade potassium nitrate. This process captures potassium perchlorate in a separate plant by treating the purge streams of the potassium nitrate re-crystallization and sodium nitrate refining processes. In separating the two sodium nitrate fractions shown in Figure 6, all the underflow streams, except perhaps some at the low temperature end of the plant, are centrifuged by means of an individual separate solid/liquid separation step. Washing is adjusted to obtain reduced perchlorate concentrations in the final product. The high perchlorate fraction is obtained after settling and centrifuging the overflow together with the lowest temperature underflow fractions.

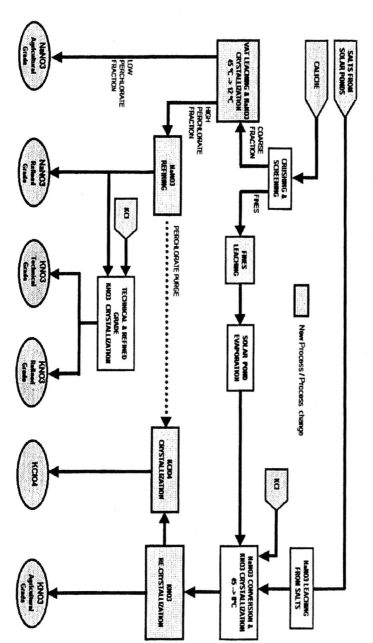

Figure 5. General Process Scheme For Low Perchlorate Production

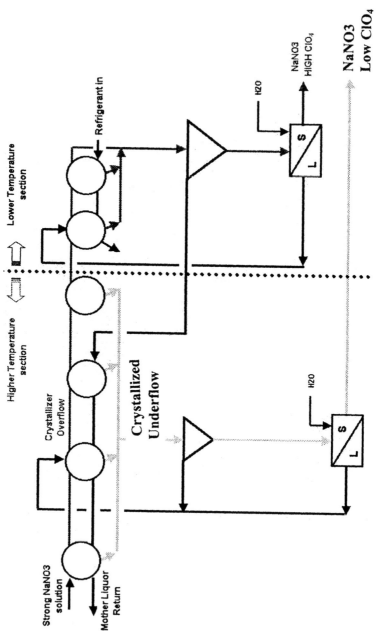

Figure 6. Proposed NaNO3 Process Scheme

The perchlorate anion is extracted from the system as $KClO_4$ by submitting the purge of the $NaNO_3$ to refining and the KNO_3 to a crystallization process at 0°C. The crystallization produces a purge (not shown in Figure 5) thus the accompanying sodium and potassium nitrates are re-circulated and recovered.

The second alternative is based on the use a selective anion exchange resin (see Figure 7), to remove perchlorate. Here re-crystallization of the agricultural grade potassium nitrate is replaced by an anion exchange process (IX ClO_4 extraction) for extracting the perchlorate from solar pond solutions containing primarily sodium nitrate (Applied Research Associates, Inc., Panama City, FL). The resin being considered for the perchlorate extraction from the brines is Purolite D 3696, a patented bi-functional anion-exchange resin with both, trihexylammonium and trieythylammonium functional groups, originally developed by Oak Ridge National Laboratory – US Department of Energy. This resin is highly selective for perchlorate, and can be regenerated displacing the perchlorate anions with tetrachloroferrate ($FeCl_4^-$) anions formed in a solution of ferric chloride and hydrochloric acid. The displaced perchlorate anion is recuperated as potassium perchlorate by reaction with potassium chloride. The degree of perchlorate extraction from the brine is adjusted to avoid saturation and precipitation of potassium perchlorate during the crystallization of potassium nitrate. This allows further refining of potassium nitrate crystals by simple washing, thus removing the need for a second crystallization process step.

CONCLUSION

SQM is committed to providing its customers with the finest fertilizer products. Upon hearing EPA's concerns with the potential health effects of perchlorate in the environment, resulting primarily from non-agricultural operations in the western U.S., and in response to customer inquiries concerning perchlorate, SQM was challenged to respond. SQM decided to conduct extensive studies which allowed the company to produce fertilizers with a significant reduction in their perchlorate content. This chapter reflects the effort SQM has made in this regard. With added expense, the company is now producing fertilizers with perchlorate levels reduced to <100 ppm. SQM has no reason to believe fertilizer use has affected in any way public health or the environment, but will continue to produce low perchlorate containing products by the most technically- and economically-sound processes currently available.

56

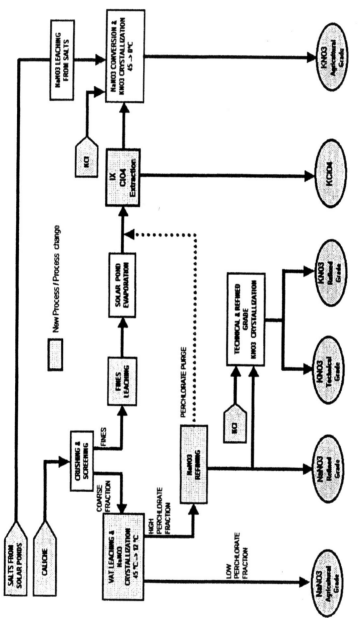

Figure 7 Proposed General Process Scheme – Ion Exchange Resin

References

1. Nelson L. B. *History of the U. S. Fertilizer Industry,* 1990 TVA
2. TFI *Annual Use of Fertilizers in the US*
3. Urbansky E. *Survey of Fertilizers and Related Materials for Perchlorate*, Final Report, EPA/600/R-01/tba, 2001
4. American Scientist, Volume 71, Number 4, July-August 1983, p. 371, cit.: Darwin 1890
5. American Scientist, Volume 71, Number 4, July-August 1983
6. SQM Internal Reports
7. *IRTC Five Year Program Plan*, June 2001, Project Proposal

Detection and Assessment of Trace Metals in Fertilizers

Chapter 5

Regulation of Heavy Metals in Fertilizer: The Current State of Analytical Methodology

Peter F. Kane[1], William L. Hall, Jr.[2], and David W. Averitt[2]

[1]Department of Biochemistry, Purdue University, West Lafayette, IN 47907-1154
[2]IMC Global, 3095 Country Road 640 West, Mulberry, FL 33860-2000

There is a trend toward regulation of non nutritive trace elements in fertilizers. The Association of American Plant Food Control Officials has developed proposed regulatory limits for As, Cd, Co, Pb, Hg, Mo, Ni, Se, and Zn. Several states are monitoring trace metals, others are considering programs. To begin evaluation of available methodology supporting such regulation, 29 labs participated in a sample exchange designed to estimate the degree of accuracy and precision possible by laboratories routinely monitoring trace metal content of fertilizer materials. Survey samples consisted of diluted solutions of certified stock standards of known concentration, and actual fertilizer materials. Laboratories used several acid digestion procedures for sample preparation, and a range of instrumentation for detection. Analytical results illustrate a lack of reasonable precision and accuracy, needed for reliable regulatory oversight. Method development activities to address these deficiencies are suggested.

Introduction

In 1997 the Seattle Times published a series of articles entitled "Fear in the Fields" (*1*) which focused the public's attention on the practice of recycling industrial waste into fertilizer products. The series raised public concern over whether potentially harmful metals could get into our soils and plants by this practice. A book entitled "Fateful Harvest" (*2*), based on the newspaper series, was published by the same author. In light of these concerns, federal and state agencies responsible for regulation of fertilizer products in the United States are evaluating risks and considering the appropriate response in the public interest.

Since there is merit in having relatively uniform rules and regulations related to fertilizers among various states, the Association of American Plant Food Control Officials (AAPFCO) early in 2001 approved Statement of Uniform Interpretation and Policy #25 (*3*). SUIP25 sets suggested upper limits of contaminant metals elements, based on a sliding scale of how much phosphorus and nutritive trace elements are claimed on the product label. Ultimately these calculations are based on formal risk assessment (*4*).

As various states consider their regulatory options, it remains to be seen how closely they might follow the SUIP25 guidelines. Whether in uniform fashion or not however, states do seem to be migrating in the direction of additional regulatory control. Currently three states are regulating various non nutritive trace elements, Washington, California, and Texas. According to a recent survey of state regulatory agencies, conducted by the authors with AAPFCO's help, 21 additional states either are, or sometime this year will be, monitoring at least some non nutritive trace elements in the fertilizer products they regulate. The states are Alabama, Arizona, Delaware, Florida, Idaho, Indiana, Kentucky, Louisiana, Maine, Maryland, Michigan, Minnesota, New Hampshire, New Jersey, New York, Oregon, Pennsylvania, South Carolina, Utah, Vermont, and Wisconsin. "Monitoring" implies analysis of fertilizers, but may or may not imply regulation based on that analysis.

This survey also asked the regulatory agencies to indicate what analytical methods are, or will be, used for existing or anticipated analysis of potentially harmful metals in fertilizers. Responses from some states were quite specific. Washington regulates 9 elements, and the methodology it specifies is EPA 3050B, and EPA 7470A/7471A for mercury. California regulates 3 elements, plus 6 more anticipated, and specifies EPA 3050B or 3051A. Texas regulates 9 elements by in house methodologies. (EPA methodologies can be found on the EPA web site at EPA.gov.) A number of other states were not as specific in designating methodologies to use however. Responses ranged from "AOAC heavy metals methods" (even though none exist for fertilizers), to "EPA", or the respondent wasn't sure, or methodologies were still to be determined.

It should be realized that there are multiple EPA methods, http://www.epa.gov/epaoswer/hazwaste/test/3xxx.htm, which can give varying results, and in addition, those methods are not intended specifically for fertilizer materials. EPA 3050B and EPA3051A are the most commonly referenced methods. EPA 3050B is a hotplate digestion with nitric acid, or sometimes nitric acid plus hydrochloric acid, depending on the elements of interest, and the type of instrumentation anticipated for use in the determination step. EPA 3051A uses a microwave rather than a hot plate. It may or may not match results from 3050B depending on digestion conditions, acids used, and the sample matrix. 3050B and 3051A were developed for use with sediment, sludge, and soil materials, not fertilizer materials, and their applicability to fertilizers has not been systematically investigated. Also, 3050B and 3051A are not intended to recover all of a given element from samples. They are intended to be leach methods to analyze samples from, for example, a superfund site. When using the methods, it is assumed that if a given environmental sample does not completely digest in the nitric acid, it is unlikely the undigested portion would have potential to leach into the ground water and escape the site.

EPA 3052, using hydrofluoric acid, is a microwave digestion procedure designed to give total element recoveries from many environmental samples. There is potentially quite a difference between leachable element content and total element content, and the magnitude of the difference could be unique to each different fertilizer sample matrix. Again, this has not systematically been investigated. There typically is a caveat in these EPA methods that says that other elements and matrices may be analyzed by the method if performance is demonstrated for the analyte of interest, in the matrices of interest, and at the concentration levels of interest. For fertilizers this has not systematically been done, so the effectiveness of the EPA methods for fertilizers is not known.

Besides variation in how samples are solublized, instrumentation commonly used for the determination step also varies. Frequently used instrumentation includes flame atomic absorption, graphite furnace atomic absorption, hydride atomic absorption, inductively coupled plasma optical emission spectrometry, and inductively coupled plasma mass spectrometry. Each class of instrumentation is subject to its own limitations, and different analysts approach these limitations with varing levels of expertise. Much remains to be done in correlating data from different instrumentation.

Given the variation in methodologies available for metals analysis, it would not be surprising if the current state of analytical agreement between laboratories doing this kind of analysis was not good. There is potential that, with 24 different state laboratories, not to mention fertilizer industry laboratories and commercial laboratories, all generating analytical data with this mix of methodologies, there could be considerable conflicting information generated. Here we report on a study designed to obtain an estimate of the degree of accuracy and precision possible by laboratories that may be asked to routinely monitor the trace metal content of fertilizer materials.

Survey Design

The survey design was limited to only 5 samples and 5 elements: As, Cd, Pb, Hg, and Se. Three of the samples were certified stock standard solutions of the 5 elements, diluted in 5% HNO_3. Solution one contained relatively low levels, representative of levels in average fertilizers. Solution two contained higher concentrations, but still below the allowed levels set by the AAPFCO SUIP25 document. Solution three contained levels above those allowed by the SUIP25 document. Analysis of these three solutions provided information on how well the laboratories were operating their various instruments, independent of digestion chemistry variation and sample matrix interference. The participating laboratories were free to use whatever instrumentation they chose.

Solution four was a real fertilizer sample, the Magruder check sample (http://www.magruderchecksample.org/) for March of 2001, predigested in nitric acid by microwave (0.25g sample, 20mL HNO_3, digested 70sec.). Analysis results for this sample included variance from matrix interferences as well as instrument variance. The fifth sample was an undigested solid fertilizer (the same Magruder check sample). The labs were instructed to digest this sample however they wished, according to their normal routine procedures. The variance associated with this sample would include instrument, sample matrix, and digestion.

A total of 29 laboratories participated in this study: 20 regulatory labs, 6 industry labs, and 3 others. The laboratories routinely analyze samples for trace element content, and would be expected to be many of the same state and industry laboratories conducting these analyses relative to regulatory activity.

Results

Table I summarizes the digestion procedures, and Table II summarizes the instrumentation used, by the 29 laboratories. Note that just because two laboratories used the same digestion equipment (microwave or hot plate) and the same digestion acid combination, this does not necessarily mean that an identical digestion procedure was employed.

Table I. Summary of Digestion Techniques Used by Laboratories

Lab	As Digestion	Cd Digestion	Pb Digestion
1	MW HNO_3	MW HNO_3	MW HNO_3
2	HP HNO_3/HCl	HP HNO_3/HCl	HP HNO_3/HCl
3	HP HNO_3/HCl	HP HNO_3/HCl	HP HNO_3/HCl
4	HP HNO_3	HP HNO_3	HP HNO_3
5		HP HNO_3/HCl	HP HNO_3/HCl
6		HP HCl	
7	HP HNO_3/HCl	HP HNO_3/HCl	HP HNO_3/HCl
8	HP HNO_3	HP HNO_3	HP HNO_3
9	MW HNO_3	MW HNO_3	MW HNO_3
10	MW HNO_3/HCl	MW HNO_3/HCl	MW HNO_3/HCl
11	HP HNO_3	HP HNO_3	HP HNO_3
12	HP Dry Ash HNO_3/HCl	HP HNO_3/H_2O_2/HCl	HP HNO_3/H_2O_2/HCl
13	HP HNO_3/HCl	HP HNO_3/HCl	HP HNO_3/HCl
14	HP HNO_3/HCl	HP HNO_3/HCl	HP HNO_3/HCl
15	HP HNO_3/$HClO_4$/HCl	HP HNO_3/$HClO_4$/HCl	HP HNO_3/$HClO_4$/HCl
16		HP HCl	
17	MW HNO_3/HCl	MW HNO_3/HCl	MW HNO_3/HCl
18	HP HCl	HP HNO_3/HCl	HP HNO_3/HCl
19		HP HNO_3	HP HNO_3
20	HP HNO_3/HCl	HP HNO_3/HCl	HP HNO_3/HCl
21	MW HNO_3	MW HNO_3	MW HNO_3
22	HP HNO_3/$HClO_4$	HP HNO_3/H_2O_2	HP HNO_3/H_2O_2
24		HP HNO_3/H_2O_2\HCl	HP HNO_3/H_2O_2/HCl
25	HP HNO_3	HP HNO_3	HP HNO_3
26	HP HCl	HP HCl	HP HCl
27	HP HNO_3/HCl	HP HNO_3/H_2O_2/HCl	HP HNO_3/H_2O_2/HCl
28	HP HNO_3/$HClO_4$	MW HNO_3	MW HNO_3
29	HP HNO_3/H_2O_2/HCl	HP HNO_3/H_2O_2/HCl	HP HNO_3/H_2O_2/HCl

(MW is Microwave, HP is Hot Plate)

Continued on next page.

Table I. Continued

Lab	Se Digestion	Hg Digestion
1	MW HNO$_3$	MW HNO$_3$
2	HP HNO$_3$/HCl	
3	HP HNO$_3$/HCl	
4	HP HNO$_3$	HP HNO$_3$
8	HP HNO$_3$	
10	MW HNO$_3$/HCl	MW HNO$_3$/HCl
11	HP HNO$_3$	HP HNO$_3$
12	HP HNO$_3$/Dry Ash/HCl	
13	HP HNO$_3$/HCl	
14	HP HNO$_3$/HCl	HP HNO$_3$/HCl
15	HP HNO$_3$/HClO$_4$/HCl	
17	MW HNO$_3$/HCl	MW HNO$_3$/HCl
18	HP HNO$_3$/HClO$_4$/HCl	HP HNO$_3$/H$_2$SO$_4$
19	HP HNO$_3$	HP HNO$_3$
20	HP HNO$_3$/HCl	
21	MW HNO$_3$	MW HNO$_3$
22	HP HNO$_3$/HClO$_4$	HP HNO$_3$/H$_2$O$_2$
25	HP HNO$_3$	HP HNO$_3$
26		HP HCl
27	HP HNO$_3$/H$_2$O$_2$/HCl	HP HNO$_3$/H$_2$O$_2$/HCl
28	HP HNO$_3$/HClO$_4$	
29	HP HNO$_3$/H$_2$O$_2$/HCl	

(MW is Microwave, HP is Hot Plate)

Table II. Summary of Instrument Techniques Used by Laboratories

Lab	As	Cd	Pb	Se	Hg
1	ICP-MS	ICP-MS	ICP-MS	ICP-MS	ICP-MS
2	ICP-OES	ICP-OES	ICP-OES	ICP-OES	
3	ICP-OES	ICP-OES	ICP-OES	ICP-OES	
4	ICP-MS	ICP-MS	ICP-MS	ICP-MS	ICP-MS
5		Flame AA	GFAA		
6		ICP-OES			
7	ICP-OES	ICP-OES	ICP-OES		
8	ICP-OES	ICP-OES	ICP-OES	ICP-OES	
9	ICP-MS	ICP-MS	ICP-MS		
10	ICP-OES	ICP-OES	ICP-OES	ICP-OES	ICP-OES
11	ICP-MS	ICP-MS	ICP-MS	ICP-MS	ICP-MS
12	AA Hydride	Flame AA	Flame AA	AA Hydride	
13	ICP-OES	ICP-OES	ICP-OES	ICP-OES	
14	ICP-OES	ICP-OES	ICP-OES	ICP-OES	ICP-OES
15	ICP-OES	ICP-OES	ICP-OES	ICP-OES	
16		Flame AA			
17	AA Hydride	Flame AA	Flame AA	AA Hydride	Cold Vapor
18	AA Hydride	Flame AA	Flame AA	AA Hydride	Cold Vapor
19		ICP-OES	ICP-OES	ICP-OES	ICP-OES
20	ICP-MS	Flame AA	ZGFAA	ZGFAA	
21	GFAA	ICP-OES	ICP-OES	ICP-OES	Cold Vapor
22	AA Hydride	GFAA	GFAA	AA Hydride	Cold Vapor
23	AA Hydride	GFAA	GFAA	AA Hydride	
24	AA Hydride	Flame AA	Flame AA	AA Hydride	Cold Vapor
25	GFAA	GFAA	GFAA	AOAC Flourometric	Cold Vapor
26	ICP-OES	ICP-OES	ICP-OES		ICP-OES
27	ICP-MS	ICP-MS	ICP-MS	ICP-MS	ICP-MS
28	AA Hydride	Flame AA	ZGFAA	AA Hydride	
29	ICP-MS	ICP-MS	ICP-MS	ICP-MS	

A statistical summary for cadmium, lead, and arsenic is given in Table III. For Cd, the three solutions prepared from accurate dilution of certified reference standards, Low, Intermediate, and High Stock Solutions, the mean of reported concentrations closely matched the True Value. In addition, the coefficients of variation (CV) for all three solutions were <15% without removal of outliers. The lowest concentration solution had the largest CV, which was expected. The data suggest that for Cd the labs were in good control of their instruments, and variation in the type of instrumentation used did not significantly affect results. The true values are not known for the material used to prepare the Predigested or Solid Fertilizer Sample, but the Cd means for the two are identical. There are four outliers, three vary by a factor of 10 or 2, raising suspicions of calculation errors. For this material it is reasonable to conclude that the sample matrix did not interfere with the analysis of Cd by the instrumentats. Difference in digestion procedures did not appear to add to the overall variance. No trends could be identified when grouping the results by instrument or digestion type.

For Pb, the means of the Low, Intermediate, and High solutions also closely matched the True Value (Table III). For the Intermediate and High solutions, the CV was similar to Cd CV's. The CV for the Low solution was much higher than the Low value for Cd, but the lead concentration in that solution was only .05ppm. The CV drops from 40 to 30 if the flame AA data is removed.

The Pb mean values for the Predigested sample and the Solid Fertilizer sample did not agree. The CV's are also larger than the corresponding CV's for the three Pb stock solutions. The CV for the Predigested sample was 122, despite the fact that three large outlier values were not included in the statistics. This strongly suggests that there are interfering elements in fertilizer matrix, resulting in one or more instruments giving erroneous results. The CV for the Pb Solid Fertilizer sample was 84, making it difficult to separate the influence of digest techniques on the total variance. Grouping the Predigested sample data and the Solid Fertilizer sample data by instrument type did not suggest any possible explanation as to sources of the interference. This may indicate that there are multiple interferences occurring among the different instrument types.

For As, (Table III), means for the stock solutions agree with the true values, but the CV's are definitely higher than the corresponding CV's for either Cd or Pb. This dispite additional outiers were excluded. There is agreement between the mean values for the Predigested and Solid samples, but the CV's are much larger than those for the stock solutions. Again, it is difficult to separate the influence of sample matrix from digestion techniques on the total variance. Uncertainty among different instrumentation types should be a concern.

Table III. Statistical Summary of Cd, Pb, and As Results (mg/kg)

Low Stock Solution	Intermediate Stock Solution	High Stock Solution	Predigested Sample	Solid Fertilizer Sample
Cadmium				
0.125[a]	0.976	1.91		
0.126[b]	0.987	1.92	44.2	44.1
0.017[c]	0.071	0.171	4.18	2.75
0.10[d]	0.88	1.50	35.3	39.7
0.18[e]	1.20	2.42	56.0	53.1
13.7[f]	7.2	8.9	9.5	6.2
None[g]	None	None	17.6, 93.8	4.3, 11
Lead				
0.05	4.88	9.52		
0.057	4.86	9.30	6.36	5.07
0.023	0.419	0.904	7.75	4.28
0.020	3.31	7.68	1.40	1.90
0.100	5.76	11.6	31.2	16.1
40.0	8.6	9.7	122	84.2
0.53, 0.26	None	0.31	66, 78, 125	37, 0.07
Arsenic				
0.05	0.976	2.86		
0.051	0.944	2.91	6.73	6.94
0.022	0.124	0.616	6.20	4.27
0.0135	0.62	2.00	0.34	0.09
0.10	1.20	4.73	26.0	19.9
42.6	13.1	21.2	92.1	61.5
0.26, 1.36	0.05, 2.2, 2.8	6.1	234, 80, 67	None

[a] True Value, [b] Mean, [c] Standard Deviation, [d] Minimum, [e] Maximum, [f] CV(%), [g] Outliers

Table IV. Statistical Summary of Se and Hg Results (mg/kg)

Low Stock Solution	Intermediate Stock Solution	High Stock Solution	Predigested Sample	Solid Fertilizer Sample
Selenium				
0.025[a]	1.95	4.76		
0.041[b]	1.82	4.46	9.40	2.24
0.019[c]	0.470	1.46	15.5	4.47
0.02[d]	0.394	0.87	0.04	0.09
0.07[e]	2.77	8.00	48.0	13.6
47.9[f]	25.8	32.9	165	200
.005, 0.2[g]	0.029	None	668	466
Mercury				
0.0025	0.049	0.48		
0.0131	0.082	0.966	3.47	0.318
0.015	0.050	0.924	3.74	0.451
0.001	0.013	0.236	0.018	0.005
0.040	0.162	3.23	8.0	0.977
113	61.5	95.7	108	142
None	None	9.47	None	None
Mercury, Cold Vapor				
0.0025	0.049	0.48		
0.0142	0.0786	0.452		
0.022	0.040	0.110		
157	51	24.2		
0.001	0.034	0.236		
0.040	0.13	0.530		
None	None	None		

[a] True Value, [b] Mean, [c] Standard Deviation, [d] Minimum, [e] Maximum, [f] CV(%), [g] Outliers

The results for Se are shown in Table IV. There is still good agreement with the true values, but the CV values are greater. The CV's for Predigested and Solid samples are large, and correlation between means is poor. It is hard to make conclusions other than to question the laboratories' ability to analyze Se.

For Hg, (Table IV), the stock solution CV's are the largest of the elements, and mean values did not match the true values, showing a positive bias. This indicates problems operating instrumentation with pure standards. The data for Predigested and Solid Fertilizers represents almost random numbers. When the cold vapor technique data are used, agreement improved for the Intermediate and High solutions, but not the Low solution. This indicates these laboratories experienced severe problems in analysis of Hg at low (trace) concentrations.

Summary

It is important to keep in mind that the survey was limited in scope. Only five elements were included in the study. And, only one real fertilizer material was involved. It is entirely possible, and expected, that other fertilizer materials would present a completely different sample matrix composition, resulting in different interfering elements, present in the digests in different proportions and concentration levels. So the particular fertilizer material used may or may not be a typical example of how other fertilizer materials might perform in a similar survey. Nevertheless, given the current state of analytical methodology, and the accuracy and precision found in this study, how well could laboratories be expected to regulate nonnutritive trace element content? Analysis of cadmium seems well behaved. It seems reasonable that different laboratories, using different techniques, could be expected to agree on the cadmium content of fertilizer materials of similar makup, and any differences between laboratories would likely be resolvable. For lead, the stock solutions were well behaved, but the real fertilizer material displayed more variability.
Laboratories might be expected to have considerable differences in the analysis of real world samples, but with collaboration and efforts to minimize instrument interferences, many such differences hopefully could be resolved. Arsenic and selenium appear considerably more difficult to analyze than lead, and laboratories might be advised to work to refine methodologies prior to use for regulation. The mercury results suggest substantial difficulties in analysis, and efforts to refine and standardize cold vapor techniques might be well advised. In an effort to inform analysts of the potential pitfalls of these analysis much of this data and information was published as paper number 16677 of the Purdue University Agricultural Experiment Station.

Suggestions for Future Work

While the negatives have been emphasized in this paper, the main positive is that twenty nine laboratories willingly participated in this study. More than a few inquired on their own initiative and asked to be included in the study. That is an indication that the laboratories recognize some of the problems inherent in the methodology, and are interested in finding real solutions. Collaborative work between laboratories is the best chance to solve some of the method difficulties and develop more robust and uniform methods of analysis.

The survey data suggests that problems caused by interfering elements in the fertilizer matrix are the major source of variation in results, and use of different digestion procedures is a minor source. Nevertheless, it is probably better to focus first on developing standard digestion procedures for fertilizers, for two reasons. First, when attempting to understand and deal with variation caused by different types of instrumentation, it is helpful if other sources of variation are minimized. Second, and more importantly, laboratories and fertilizer regulators have not yet adequately dealt with the fact that the most commonly used digestion procedures are leach procedures, rather than procedures designed for total recovery of the elements present in the sample matrix.

The most commonly used versions of those leach procedures, EPA 3050 and 3051, were not designed for fertilizer materials. We do not now know whether typical fertilizer materials, digested according to EPA 3050 or 3051, would result in digestates with a high percentage of the elements of interest in solution, or would result in substantially less than total solubilities of those elements. It is our intention to analyze a range of representative fertilizer material types using EPA 3051 and EPA 3052 as the digestion procedures. We will focus exclusively on microwave, rather than hot plate digestions, because by nature the microwave temperature and pressure parameters are more reproducibly controlled, the vessels are sealed, controlling volatization loss of certain elements. We will also investigate the effect of increasing the temperature/pressure/digestion time parameters of EPA 3051 to see if this gives increased correlation to results from the EPA3052 procedure. It is anticipated that the combination of a stable digestion procedure and analysis by a high resolution ICP-MS instrument capable of better differentiation of elements of interest from other interfering elements present in the digests, will result in definitive analytical results which can be used as reference points for calibration of other instrumentation.

Survey Participants

1. Dean Abrams, Michael Hojjatie, Tessenderlo Kerley, Inc, 2480 West Twin Buttes Road, Sahuarita, AZ 85629.
2. Dean Alcorn, John Kennedy, Agrium, 3010 Conda Road, Soda Springs, ID 83276.
3. David W. Averitt, Charles N. Kinsey, IMC Phosphates Company, P.O. Box 2000, Mulberry, FL 33860-1100.
4. David Boggeman, Montana State University, A.E.S. Analytical Laboratory, Bozeman, MT 59717-3620.
5. Ann Renea Brinlee, Oregon Department of Agriculture, Laboratory Services, Portland, OR 97209.
6. Michelle Campbell, Zena Kassa, Minnesota Department of Agriculture, Laboratory Services Division, St. Paul, MN 55107.
7. Mario Dupuis, Canadian Food Inspection Agency, 960 Carling Avenue, Ottawa, Ontario, Canada K1AOC6.
8. Marc E. Engel, Florida Department of Agriculture, 3125 Conner Blvd. #9, Tallahassee, FL 32399.
9. Harold Falls, Sanford Siegel, John Longest, Barton Boggs, CF Industries, Inc., 10608 Paul Buckman Highway, Plant City, FL 33565.
10. Julia Gantchev, Doug Marsh, Arizona Department of Agriculture, 2422 West Holly Street, Phoenix, AZ 85009.
11. John Harriger, California Department of Food and Agriculture, Feed Laboratory, 3292 Meadowview Road, Sacramento, CA 95832.
12. Cham Hoang, Utah Department of Agriculture and Food, 350 North Redwood Road, Salt Lake City, UT 84114-6500.
13. David E. Lichtenberg, University of Kentucky, Division of Regulatory Services, Lexington, KY 40546-0275
14. Don Jernstrom, Angela H. Nguyen, Cargill Fertilizer, Inc., 8813 Highway 41S, Riverview, FL 33569.
15. Mark Lee, California Department of Agriculture, 3292 Meadowview Road, Sacramento, CA 95832.
16. Steve McGeehan, Idaho Department of Food Science and Toxicology, Analytical Sciences, Moscow, ID 83844-2203.
17. Craig Musante, Connecticut Agricultural Experiment Station, P.O. Box 1106, New Haven, CT 06511-2016.
18. Natalie Newlon, Office of the Indiana State Chemist, Purdue University, West Lafayette, IN 47907-1154.
19. Offiah Offiah, Maryland Department of Agriculture, State Chemist Section, 50 Harry S. Truman Parkway, Annapolis, MD 21401.

20. Jayesh Pathak, Pennsylvania Department of Agriculture, 2301 North Cameron Street, Harrisburg, PA 17110-9408.
21. Wayne Robarge, North Carolina State University, 100 Derieux Street, Raleigh, NC 27695-7619.
22. Joanne Steffes, Agrium Redwater Fertilizer Complex, Bag 20, Redwater, Alberta, Canada TOA 2WO.
23. Don Tate, Connie Zmrhal, Illinois Department of Agriculture, Chemistry Laboratory, P.O. Box 19281, Springfield, IL 62794-9281.
24. Laure Taylor, Thornton Laboratories, 1145 East Cass Street, Tampa, FL 33602.
25. Nancy Thiex, Terri VanErem, Nancy Anderson, and Renata Wnuk, Olson Biochemistry Laboratories, South Dakota State University, Brookings, SD 57007-1217.
26. Argentina Vindiola, Aria McCall, James Embry, Office of Texas State Chemist, P.O. Box 3160, College Station, TX 77841- 3160.
27. Cindy Wagner-Wiebeck, Nebraska Department of Agriculture, 3703 South 14th Street, Lincoln, NE 68502-5399.
28. David Wall, LSU Agricultural Center, P.O. Box 25060, Baton Rouge, LA 70803.
29. Gordon Wallace, Perkin Elmer, 510 Guthridge Court, Norcross, GA 30092.

References

1. Wilson, D. *Fear in the Fields: How Hazardous Wastes Become Fertilizer, Part 1*; The Seattle Times Company, P.O. Box 70, Seattle, WA 98111, July 3, 1997.
2. Wilson, D. *Fateful Harvest;* HarperCollins: New York, NY, 2001.
3. *Statement of Uniform Interpretation and Policy #25*; Association of American Plant Food Control Officials, D. L. Terry, Secretary, University of Kentucky, 103 Regulatory Services Bldg., Lexington, KY 40546-0275.
4. *Scientific Basis for Risk-Based Acceptable Concentrations of Metals in Fertilizers and Their Applicability as Standards*; The Weinberg Group Inc., 1220 Nineteenth St. NW, Suite 300, Washington, DC 20036-2400.

Chapter 6

Determination of Trace Metal Content of Fertilizer Source Materials Produced in North America

Wayne P. Robarge[1], Dennis Boos[2], and Charles Proctor[2]

Departments of [1]Soil Science and [2]Statistics, North Carolina State University, Raleigh, NC 27695

There is increasing concern over the trace metal content of fertilizers and their subsequent application to agricultural and urban lands. It is feared that continued addition of trace metals to soils via fertilizers poses potential risks to farm families and to consumers of farm products. In order to assess this potential risk, it is necessary to know the trace metal content of fertilizer source materials. This study was undertaken to generate a statistically valid sampling of fertilizer source material produced in North America, and to determine the trace metal content (As, Cd, Cr, Cu, Mo, Ni, Pb, Se, V, U and Zn) of the resulting composite samples generated by the sampling protocol using modern analytical instrumentation and accepted good laboratory practices. The results support the general hypothesis that phosphate bearing source materials do contain varying levels of trace metals. The results also demonstrate that non-phosphate bearing N-P-K source materials do not contain significant amounts of trace metals and should not be considered significant sources of metals when added to agricultural or urban soils. Agreement of analyses from an interlaboratory comparison demonstrate the suitability of the analytical protocols used in this study.

© 2004 American Chemical Society

Introduction

Fertilizer source materials are known to contain trace metals in varying amounts. There is increasing concern over the trace metal content of fertilizers and their subsequent application to agricultural and urban lands. It is feared that the continued addition of trace metals to soils via fertilizers poses potential risks to farm families and to consumers of farm products. Even though the U.S. EPA has released their own study (1) declaring most fertilizers not a concern in regards to exposure to trace metals, actions to regulate metal contents at the state level continue, such as the states of Washington, Texas, California and Oregon, Michigan and Maine. How these states formulate regulatory policy regarding trace metals in fertilizers may well portend similar actions among the remaining states in the nation.

It is generally acknowledged that phosphate-bearing fertilizer source materials are the source of most trace metals associated with fertilizers (2,3,4), excluding fertilizer materials specifically formulated to contain micronutrients. However, while there have been a number of published studies on the chemical analysis of fertilizers, often the source and therefore representativeness of the actual fertilizer materials reported on has been generally poorly documented. In addition, there has also been growing concern over the fact that there are no standard analytical protocols available for determining the trace metal content of fertilizer source materials and fertilizer blends (5). The range in composition of fertilizer source materials and fertilizer blends poses a number of severe challenges in the successful determination of their trace metal content. The extent and nature of these challenges is just now being better understood (5).

Given the uncertainty surrounding the trace metal content of fertilizers, in particular fertilizer source materials, this study was designed to address three main objectives: (1) generate a statistically valid sampling of fertilizer source material production facilities in North America, (2) generate composite samples representative of the different fertilizer source materials, and (3) conduct the chemical analysis of the composite samples using modern chemical instrumentation and accepted good laboratory practices. Presented here are results for the As, Cd, Cr, Cu, Mo, Ni, Pb, Se, V, U and Zn content of several sampled fertilizer source materials (Table I) sampled over a two year period (1999 - 2001).

Materials and Methods

Sampling Scheme and Sample Handling

The statistically based sampling scheme used in this study was centered on inductive statistical procedures, which assumed the existence of a population that can be characterized through the selection of random samples. The mean value of the population is deemed important because fertilizer source materials are mainly dispersed in bulk, not as individual units. Focusing on the means of the respective populations also aids the actual collection of samples and sample processing by allowing for compositing, which in turn reduces the analytical burden associated with the project.

Table I. Sampled fertilizer source materials.

Designation	Source Material
DAP	Di-Ammonium Phosphate
MAP	Mono-Ammonium Phosphate
AS	Ammonium Sulfate
KCl	Potassium Chloride
TSP	Triple Super Phosphate
UREA	Urea
AN	Ammonium Nitrate
SPM	Sulfate of Potash Magnesium

The population to be characterized for a given fertilizer source material was defined as the total production of product from all The Fertilizer Institute (TFI; http:www.tfi.org) member fertilizer source material production facilities within a given time period (typically 30 days). For example, for TFI member facilities producing DAP/MAP, the population to be characterized would be the amount of DAP/MAP produced in North America during a given 30 day period. This afforded the study a number of advantages in arranging for sample collection. First, the number of times (days) a given facility would be asked to submit samples during a 30 day period would be a function of the tonnage of product produced by the facility compared to the total produced by all facilities during the designated 30 day period. Second, there would be no need for alternative sampling instructions should a facility not be operating on the day selected for sampling, nor if the plant experienced an unexpected shutdown during the day selected for sampling that location. If a production facility were not operating it could not contribute to the total population produced by the TFI member facilities for a given product during the designated 30-day period.

Lastly, there would be no need for special handling or compositing of samples at each production facility. All that would be required would be representative samples of the product obtained during the 24-hour period. This could easily be accomplished using established protocols for obtaining quality control samples at each production facility, minimizing the amount of effort required by each production facility to actually participate in the study. Based on results from a preliminary sampling study (data not shown), it was determined that 4 samples (0.5 to 2 kg ea) obtained 6 hours apart were sufficient to characterize the product generated during a 24 hour period. Thus the maximum number of samples that could have been obtained during a given 30-day period is 120 (30 sampling days * 4 samples from a designated plant per day = 120).

The composite samples for chemical analysis were generated at North Carolina State University (NCSU) after receipt of all the samples collected for a given fertilizer source material during a given designated period. Using a randomized statistical design, the individual samples were reduced using a mechanical rotary splitter (Gilson Company, Inc., Model SP-201V). The eight composite samples were based on first numbering production facilities geographically across North America and then differentiating them by even-numbered versus odd-numbered production sites. Then the samples were randomized into four groups corresponding to the 4 sampling times specified for each day. This scheme resulted in the division of the sampling bags to form the 8 composite samples [2 (odd versus even) times 4 sampling times per day = 8 composite samples per sampling period]. The 120 bags of material were put, at most, twice through a rotary splitter with 16 compartments so as to obtain, by random allocation, a save amount targeted at 18 grams. Each composite thus should have attained 270 grams in mass. In fact, less than the target amount of sample bags were collected due to non-responses. For the majority of fertilizer source materials sampled, the response rate was closer to 80% (Table IV), lowering the composite sample mass to approximately 200 grams.

No specific adjustment was made for non-responses during a sampling period. It was considered that such cases happened "at random." The sampling rates calculated for each sampling bag from the production line tons per hour to the grams of composite were found to have a coefficient of variation (CV) = 61%. In instructions sent to each production facility there were rough guidelines for the desired amounts to be collected which were designed to achieve a CV = 68%. Our calculations indicate that the local operators who actually obtained the samples were more careful than our target guidelines.

The final step in the preparation of the composite samples was to convert them to a uniform size using a Brinkman ZM100 stainless steel mill equipped with a 0.5 mm stainless steel screen. This resulted in the bulk of the composite samples having a particle size of between 100 - 250 microns.

Sample Digestion and Chemical Analysis

Two different wet digestion techniques were used to prepare the ground composite samples for analysis. The selection of the digestion techniques was predicated on the following assumptions: (a) most fertilizer source materials are in fact pure salts which do not require aggressive digestion techniques to obtain sample dissolution; (b) many of the sampled fertilizer source materials will in fact contain only trace levels of metals, therefore the digestion protocol should minimize potential sample contamination; and (c) a digestion protocol that allows for relatively large sample sizes (0.5 to 1.0 grams) is preferred to minimize large dilution factors in order to achieve as low detection limits in the final solids as possible. It would also be beneficial to minimize the volume of concentrated acids required to achieve sample dissolution to avoid possible systematic errors known to occur with sample introduction systems for most ion-coupled plasma (ICP)-emission spectrometers (ICP-AES) or ICP-mass spectrometers (ICP-MS) when using relatively concentrated acid solutions (>0.5 molar) (6,7,8).

The wet digestion technique selected for the DAP/MAP, TSP, and SPM composites was a microwave Parr bomb equipped with a 25mL Teflon crucible. Typically, 0.1 to 0.25 grams of sample was digested using 5mL of ultra-pure HNO_3. Addition of 70uL of HF with the HNO_3 was sometimes used to facilitate dissolution of samples containing silicates or relatively high concentrations of titanium. The heating cycle was carried out in a standard commercial microwave oven (600 Watt rating) equipped with a rotating carousel. Complete sample dissolution was obtained using a Power setting of 30 for a total of 25 minutes. (Note: A 250 mL Teflon bottle filled with 200 mL distilled water placed in the center of the Parr bombs is necessary to balance microwave loading during heating.) At the end of the cycle, the Parr bombs (5 per cycle) were removed and allowed to cool for one hour before opening. Following cooling the contents were quantitatively transferred to a polystyrene 50mL centrifuge tube. Final solution volume was set at 30mL resulting in an acid strength of approximately 2.5M HNO_3, and dilution factors ranging from 120 to 300 depending on initial sample mass.

For AS, KCl, UREA and AN, an acid reflux digestion technique was selected using 125 or 250mL Teflon bottles (FEP-C or PFA) with Teflon screw caps. Typically, 0.5 to 1.0 grams of sample was digested using 5mL of concentrated ultra-pure HNO_3. Addition of concentrated HF or HCl with the HNO_3 was sometimes used as well to facilitate dissolution of samples. Typically the samples were allowed to stand overnight at room temperature before heating. The heating cycle was carried out in a standard commercial microwave oven (1000 Watt rating) with the Teflon bottles placed inside of a covered modified plastic container (microwave compatible) through which was passed 100% di-nitrogen gas (N_2) at a rate of 2L per min. As with the Parr bombs, a Teflon bottle containing 200mL of deionized water was included during each heating cycle to buffer the intensity of microwaves within the oven cavity. Typically the heating cycle consisted of a total of 30 minutes with alternating 5-minute cycles of heating (Power level set at 5%) and standing. Between 5 to 8 digestion bottles could be included with each run. The bottles were capped, with the caps adjusted to finger tightness and then turned counterclockwise an eighth of a turn. This prevented undo pressure buildup during the digestion step, but also allowed for refluxing of acid within the digestion bottles. The use of the covered modified plastic container with flowing N_2 gas kept acid fumes from damaging the interior of the oven without the need for making modifications to accommodate a special exhaust system. After cooling, the contents were quantitatively transferred to a polystyrene 50mL centrifuge tube. Final solution volume was set at 45mL resulting in an acid strength of approximately 1.5M

HNO$_3$, and dilution factors ranging from 45 to 90 depending on initial sample mass.

The elemental content of the HNO$_3$ digestates was determined using ICP-AES in axial view mode (Perkin Elmer Model 2000DV) (Table II). An internal standard was not used. Previous work had demonstrated that use of the cyclonic spray chamber with the concentric nebulizer eliminated signal suppression due to matrix effects that are common with cross-flow nebulizers (9). In addition, reduction of the nebulizer gas flow rate was used to decrease solvent and matrix plasma loading, with the additional benefit of increasing aerosol residence time and the efficiency of energy transfer to the analyte (10). Robustness of the plasma was also monitored by measuring the Mg II 280.270 nm to Mg I 285.213 nm line intensity ratio (11). Only acid digestates generated from the KCl source material demonstrated measureable reductions in signals when sample amounts of 1 gram were used. All standards were prepared from ICP-grade National Institute of Standards and Technology (NIST) traceable stock solutions (SPEX CertiPrep or GFS Chemicals) in 2.5M ultra-pure HNO$_3$. No attempt was made to match the standard matrices with sample matrices in terms of salt content. Analytical wavelengths and calculated method detection limits are listed in Table III.

Table II. Operating conditions for ICP-emission spectrometer.

Item	*Settings*
Chamber	Unbaffled Cyclonic (quartz)
Nebulizer	Concentric (quartz, high solids)
Power	1500 Watts
Plasma Flow	17 L/min
Auxillary Flow	0.2 L/min
Nebulizer Flow	0.58 L/min
Pump Rate	1.1 mL/min
Heat	30 - 33 C
Delay	90 sec
Integration	2 - 10 sec
Purge (N2)	normal

InterLaboratory Analytical Comparison

Subsamples from the 16 composites from the majority of fertilizer source materials sampled were sent to three laboratories for additional chemical analyses: Nuclear Services, Department of Nuclear Engineering, NCSU (Mr. Scott Lassell, Manager; Instrumental Neutron Activation Analysis - INAA); CF Industries, Zephyrhills, FL (Contact: Mr. Harold Falls, hot-plate digestion, ICP-AES); and IMC-Global, Mulberry, FL (Contacts Mr. Bill Hall and Mr. Dave Averitt, microwave-digestion, ICP-MS). Analysis via INAA provided an estimate of the elemental content of the samples without the need for sample dissolution. The analyses provided by CF Industries and IMC-Global provided comparisons between analytical instrumentation and digestion protocols.

Results and Discussion

Sampling periods and sampling targets for each fertilizer source material are provided in Table IV. For DAP, MAP, AS, KCl, and TSP, the average sample response was approximately 80%, which was deemed highly acceptable for obtaining a representative sampling from these production facilities. Sample response was consistently 63% for Urea and dropped to 49% for AN. The low response for AN was due to a rapid increase in the price of natural gas (a feed stock in the production of AN) during the sampling period. Many plants ceased operation due to high production costs. The low recovery for Urea may reflect a yearly slow down in production due to a decrease in market demand during the sampling periods selected. SPM sample numbers were reduced as only one TFI-member site produces this material in North America. TSP sample numbers were reduced because only several sites produce this material on an intermittent basis based on demand and in limited amounts compared to the production of DAP/MAP.

Table III. Analytical wavelengths and calculated method detection limits.

Element	Wavelength	$LLOD^a$	RSD^b	MDL^c	MDL
	- nm -	- mg/L -	- % -	- mg/kg -	- mg/kg -
As	193.696	0.010	9.7	1.0	0.3
Cd	214.440	0.010	0.7	0.04	0.01
Cr	267.716	0.025	1.5	0.2	0.01
Cu	327.393	0.010	1.1	0.06	0.02
Mo	202.031	0.025	1.9	0.3	0.02
Ni	231.604	0.010	16.6	1.0	0.3
Pb	220.353	0.010	3.5	1.0	0.4
Se	196.026	0.010	18.6	1.0	0.3
V	311.071	0.025	1.8	0.05	0.01
U	409.014	0.025	8.8	1.0	0.3
Zn	206.200	0.010	1.4	0.08	0.02

a. Selected lower limit of detection.
b. Relative standard deviation (n = 8 replicates).
c. Method detection limit, sample mass 0.17 or 1.0 grams.

Table IV. Sampling schedule for fertilizer source materials.

Product	Dates	Target	Sample Collection Actual	% Actual
DAP/MAP	Feb.-Mar. 99	120	84	70
AS	April-May 99	120	104	87
KCl	May-June 99	120	95	79
UREA	Aug.-Sept. 99	120	74	62
SPM	Mar.-April 00	12	10	83
DAP/MAP	Mar.-April 00	120	91	76
TSP	Mar.-April 00	60	45	75
AS	Mar.-April 00	120	96	80
AN	April-May 00	120	59	49
UREA	May-June 00	120	76	63
KCl	April-May 01	120	104	87

DAP/MAP and TSP

Except for Se, and for Pb in DAP/MAP, quantitative amounts of the trace metals were determined in the composite samples for DAP/MAP and TSP (Table V). An estimate of temporal and site-to-site variability (differences in final product composition that may be due to variation in the trace metal composition of the parent ore) is available from the data for DAP/MAP. Mean As values were essentially identical for the DAP/MAP composites with CV <15%. Note as well that the %CV for the single sampling of TSP is only 13%. Other trace metals with relatively low %CV are Mo and U. All three of these trace metals probably occur as anions in the fertilizer matrix. The data in Table V implies that there is relatively little temporal or site-to-site variability in the mean composition of As, Mo and U in DAP/MAP produced in North America across the 2-year sampling period for this study.

There is substantially more variation in the other elements. It is likely that site-to-site variation (geological variation in the trace metal composition of the parent ore) is the primary source of the observed differences in mean values between samplings, although temporal variation within a site cannot be discounted. The calculated %CV values range from 15 to >35% and illustrate the range in uncertainty for these elements likely in fertilizer blends that would be formulated from different sources of DAP/MAP or TSP sampled in this study. However, despite the degree of variability exhibited for DAP/MAP, the mean values for the composites for the two sampling periods essentially agree within a standard deviation.

Table V. Trace metal content of DAP/MAP and TSP[a].

Element	DAP/MAP Feb. - Mar. 99 Mean	SD	DAP/MAP Mar. - April 00 Mean	SD	TSP Mar. - April 00 Mean	SD
As	11.9	1.7	11.4	1.3	9.2	1.2
Cd	15.8	2.8	22.0	8.9	19	15
Cr	152	36	141	31	115	41
Cu	7.4	3.5	6.5	3.0	8.2	3.4
Mo	10.2	1.4	9.0	0.5	7.5	0.6
Ni	29	4	33	9	39	18
Pb	< 1	-	< 1	-	5.2	0.6
Se	< 1	-	< 1	-	< 1	-
V	169	3	191	47	193	68
U	161	8	165	7	143	16
Zn	188	54	244	88	230	158

a. Units=mg/kg; n = 8; SD = one standard deviation.

KCl, UREA, AN, AS, and SPM

Only quantitative amounts of Cr, Ni and Zn were determined for AS, SPM (Table VI), KCl (Table VII), and UREA (Table VIII). Levels of As, Cd, and Pb, using the protocols described in this text were all below detection limits. The amounts of Cr, Ni and Zn detected in AS, KCl and UREA were essentially <1 mg/kg with %CV values ranging from 15 to 100%. This strongly suggests that these three elements are present because of contamination associated with sample handling, either at the respective production facilities or more likely during the particle size reduction step used to prepare the composite samples for analysis. Indeed, the ratio of Cr to Ni for AS in Table VI is essentially constant between the composites generated from samples collected in 1999 and 2000. The degree of Cr contamination from sample handling should also be a function of source material hardness. The mean Cr values for AS were 0.52 and 1.2 mg/kg. For KCl the corresponding values were 0.13 and 0.27 mg/kg, and for UREA were 0.25 mg/kg. A somewhat similar trend was observed for Ni. The only material not passed through the mill was AN. Measured concentrations for Cr are 0.045 mg/kg (Table VIII) reinforcing the hypothesis that the Cr and Ni concentrations observed were primarily due to sample preparation during the particle size reduction step.

There were no consistent trends observed for Zn among AS, KCl or UREA but it is hypothesized that the source of the variation in measured Zn concentrations is due to contamination during sample handling or product formulation. In any event, the mean concentration of all elements was < 1 mg/kg, reinforcing the notion that these fertilizer source materials are not significant sources of trace metals.

Table VI. Trace metal content of AS and SPM[a].

Element	AS April - May 99 Mean	SD	AS Mar. - April 00 Mean	SD	SPM Mar. - April 00 Mean	SD
As	< 0.3	-	< 0.3	-	< 1	-
Cd	< 0.01	-	< 0.01	-	< 0.04	-
Cr	0.52	0.07	1.19	0.39	42	22
Cu	< 0.02	-	< 0.02	-	0.5	0.3
Mo	< 0.02	-	< 0.02	-	0.5	0.2
Ni	0.31	0.27	0.51	0.16	18	10
Pb	< 0.4	-	< 0.4	-	< 0.4	-
Se	< 0.3	-	< 0.3	-	< 1	-
V	0.05	0.02	0.07	0.09	< 0.05	-
U	< 0.3	-	< 0.3	-	< 1	-
Zn	0.56	0.26	0.33	0.25	0.4	0.5

a. Units = mg/kg; n = 8; SD = one standard deviation.

Table VII. Trace metal content of KCl[a].

Element	May - June 99 Mean	SD	April - May 01 Mean	SD
As	< 0.3	-	< 0.3	-
Cd	0.06	0.01	0.04	0.02
Cr	0.13	0.09	0.27	0.10
Cu	0.42	0.01	0.25	0.05
Mo	< 0.02	-	< 0.02	-
Ni	0.31	0.09	0.30	0.01
Pb	< 0.4	-	< 0.4	-
Se	< 0.3	-	< 0.3	-
V	< 0.01	-	< 0.01	-
U	< 0.3	-	< 0.3	-
Zn	0.4	0.8	0.4	0.5

a. Units = mg/kg; n = 8; SD = one standard deviation.

The amounts of Cr and Ni found associated with SPM (Table VI) are comparable to amounts found for DAP/MAP and TSP (Table V). This suggests that SPM as formulated contains significant amounts of Cr and Ni, although contamination during sample handling or product formulation from grinding cannot be discounted due to the inherent hardness of this source material. Indeed, trace metal analyses for the SPM product are available via the Internet (http://www.regulatory-info-imc.com). Reported Ni concentrations are only 1 to 2 mg/kg which is substantially less than the value of 18 mg/kg (Table VI). Sample preparation for the posted analyses of the SPM product avoids the use of stainless steel mills (Mr. Bill Hall, IMC-Global, personal communication), supporting the conclusion that the Cr and Ni concentrations found in SPM were derived primarily from contact with the stainless steel mill during sample preparation.

Table VIII. Trace metal content of UREA and AN[a].

| | UREA | | | | AN | |
| | Aug. - Sept. 99 | | May - June 00 | | April - May 00 | |
Element	Mean	SD	Mean	SD	Mean	SD
As	< 0.3	-	< 0.3	-	< 0.3	-
Cd	< 0.01	-	< 0.01	-	< 0.01	-
Cr	0.25	0.04	0.25	0.06	0.045	0.006
Cu	0.05	0.03	< 0.02	-	< 0.02	-
Mo	< 0.02	-	< 0.02	-	< 0.02	-
Ni	0.29	0.05	0.32	0.08	< 0.3	-
Pb	< 0.4	-	< 0.4	-	< 0.4	-
Se	< 0.3	-	< 0.3	-	< 0.3	-
V	< 0.01	-	< 0.01	-	< 0.01	-
U	< 0.3	-	< 0.3	-	< 0.3	-
Zn	0.07	0.03	0.24	0.08	< 0.02	-

a. Units = mg/kg; n = 8; SD = one standard deviation.

Interlaboratory Analytical Comparison

Results for the composite DAP/MAP samples generated from the first sampling period are provided in Table IX. With DAP/MAP, very good agreement was observed for As, Cd, Se and V. The remaining elements agreed within an order of magnitude, but there appear to be some potential trends in the data. For example, the results obtained for Cr, U and Ni are consistently higher than those reported by the other three laboratories, while no Pb was detected compared to the mean values of 4 and 2 reported using ICP-MS and ICP-AES, respectively (Table IX). These results suggest the presence of systematic bias in the analyses either in the protcols used in this study and/or one or more of the participating laboratories. The results for Cu, Mo and Zn are inconsistent in that there is either no close agreement, or only agreement between pairs of participating laboratories.

Results for the composite KCl and AS samples generated from the first sampling period are provided in Table X. Overall there is good agreement between the laboratories in that the majority of results are at detection limits, the estimates of which vary among protocols. There are, however, some notable exceptions. For KCl, analyses performed by ICP-MS reported substantial values for Cr, Cu, Ni, Pb and Zn, while the other laboratories reporting data cited primarily detection limits or much lower concentrations for these elements. A similar trend is observed for AS (Table X) in that a large number of the results are at detection limits, but analyses performed by ICP-MS report relatively substantial values for Cu, Ni, Pb and Zn. The reasons for these discrepancies are not known. KCl and AS represent challenging sample matrices in that they are essentially pure salts and may cause spectral or mass interferences not typically encountered. There are also differences in digestion procedures between the laboratories, suggesting lack of complete recovery in some instances, however, such a conclusion is not consistent with the results in Table IX. In general, the intercomparison data supports the conclusion that the analytical protocols adopted for use in this study are appropriate and capable of producing accurate and precise results. This intercomparison data also illustrates the degree of agreement between different analytical laboratories that is possible using differing analytical instrumentation and digestion techniques for uniformly prepared sample materials containing trace metal compositions substantially above detection limits. However, it should also be recognized that the laboratories selected for the interlaboratory analytical comparison were experienced in dealing with sample matrices as potentially complicated as fertilizer source materials.

Summary

A statistically valid sampling of TFI member fertilizer source material production facilities in North America to generate composite samples representative of the fertilizer source materials DAP/MAP, TSP, AS, KCl, UREA, AN and SPM has been completed. Chemical analyses of the composite samples was carried out using modern chemical instrumentation and accepted good laboratory practices. The data generated from the study are consistent with the general hypothesis that phosphate bearing source materials do contain varying levels of trace metals, while source materials such as AS, KCl, UREA, AN do not and should not be considered significant sources of metals when added to agricultural or urban soils. SPM was also found not to contain trace metals except for Cr and Ni. The presence of significant amounts of Cr and Ni in the SPM product is best explained as contamination introduced during the particle size reduction step used to generate the composite samples (stainless steel mill equipped with a stainless steel screen) due to the inherent hardness of this source material. Results from analyses of splits from the original composite samples by three external laboratories using a variety of digestion protocols and analytical instrumentation in general support the conclusion that the analytical protocols adopted for use in this study are appropriate and capable of producing accurate and precise results. However, there were significant deviations noted, suggesting continued refinement of the analytical protocols adopted for use in this study is required for certain elements (e.g. Ni and Pb). Agreement with results from the external laboratories was more problematic near detection limits, and varied with sample matrix and element. This observation is not inconsistent with the results of other intercomparison studies for trace metals (5) and illustrates the continuing need to develop and refine analytical protocols in order to accurately and precisely determine the trace metal content in fertilizer source materials and fertilizer blends.

Table IX. Intercomparison analyses for DAP/MAP (Feb.-Mar. 99)[a].

Element	NCSU INAA	IMC-Global ICP-MS	CF Industries ICP-AES	This Study ICP-AES
As	11.5	11.1	11.8	11.9
Cd	< 15	16.0	14.2	15.6
Cr	128	136	136	152
Cu	-	9.8	6.5	7.4
Mo	-	13.5	10.5	10.2
Ni	< 30	26	23	29
Pb	-	4.4	2.2	< 0.4
Se	< 1.2	< 1.7	0.7	< 1
V	173	159	167	169
U	147	151	157	161
Zn	153	184	145	188

a. Mean values; n = 8; units = mg/kg.

Acknowledgements

The support, encouragement and patience of Mr. Bill Herz, Project Manager, and The Fertilizer Institute are hereby duly acknowledged in the execution of this study. A special thank you is extended to Mr. Bill Hall, Mr. Dave Averitt of IMC-Global and Mr. Harold Falls, CF Industries, for their

Table X. Intercomparison analyses for KCl and AS^a.

Element	NCSU INAA	IMC-Global ICP-MS	CF Industries ICP-AES	This Study ICP-AES
		--KCl (May - June 99)--		
As	< 2	< 0.1	-	< 0.3
Cd	< 14	< 0.08	-	0.06
Cr	< 1.1	1.8	-	0.13
Cu	-	19	-	0.42
Mo	< 8	< 0.1	-	< 0.02
Ni	< 9	12	-	0.31
Pb	-	4.9	-	< 0.4
Se	< 0.9	< 0.5	-	< 0.3
V	-	< 0.1	-	< 0.01
U	< 2	< 0.2	-	< 0.3
Zn	< 5	30	-	0.4
		--AS (April - May 99)--		
As	< 0.03	< 0.1	< 0.1	< 0.3
Cd	< 0.4	< 0.08	< 0.1	< 0.1
Cr	0.9	1.1	1.0	0.52
Cu	-	4.8	0.3	< 0.02
Mo	< 0.6	0.14	0.09	0.41
Ni	< 1.8	2.5	0.60	< 0.3
Pb	-	2.9	< 1	< 0.4
Se	< 0.3	< 1.6	0.9	< 0.3
V	-	< 0.1	< 0.1	0.05
U	< 0.05	< 0.03	< 0.1	< 0.3
Zn	-	8.5	2.7	0.56

a. Mean values; n = 8; units = mg/kg.

contributions to this important project. Lastly, acknowledgement is hereby given to the plant operators/lab managers of TFI fertilizer source material production facilities in North America for their assistance in gathering the many samples needed to satisfy the demands of the project's statisticians.

References

1. *Background Report on Fertilizer Use, Contaminants and Regulations.* National Program Chemicals Division. Office of Pollution Prevention and Toxics. U.S. Environmental Protection Agency, Wachington, D.C. 20460. Contract No. 68-D5-0008.
2. Raven, K.P.; Loeppert, R.H. *Commun. Soil Sci. Plant Anal.* 1996, 27, 2947-2971.
3. Raven, K.P.; Reynolds, J.W.; Loeppert, R.H. *Commun. Soil. Sci. Plant Anal.* 1997, 28, 237-257.
4. Bashour, I.; Hannoush, G.; Kawar, N. In *Environmental Impact of Fertilizer on Soil, Air and Water*; Hall, W.; Robarge, W.P.,Eds.; ACS Sympoisum Series; American Chemical Society:Washington D.C., 2003 (in press).
5. Kane, P.F.; Hall, W.L.; Averitt, D.W. In *Environmental Impact of Fertilizer on Soil, Air and Water*; Hall, W.; Robarge, W.P.,Eds.; ACS Sympoisum Series; American Chemical Society:Washington D.C., 2003 (in press).
6. Canals, A.; Gras, L.; Contreras, H. *J. Anal. At. Spectrom.* 2002, 17, 219-226.
7. Todoli, J.L.; Mermot, J.M. *Spectrochimica Acta Part B.* 1999, 54, 895-929.
8. Todoli, J.L.; Mermot, J.M. *J. Anal. At. Spectrom.* 2000, 15, 863-867.
9. Robarge, W.P.; Falls, H.; Hall, W.; Averitt, D.; Kane, P. *222nd American Chemical Society National Meeting, Division of Agrochemicals*, Session: Environmental Impact of Fertilizer Products on Soil, Air, and Water. August 25-29, 2001. Chicago, IL.
10. Todoli, J.L.; Gras, L.; Hernandis, V.; Mora, J. *J. Anal. At. Spectrom.* 2002, 17, 142-169.
11. Grotti, M.; Magi, E.; Frache, R. *J. Anal. At. Spectrom.* 2000, 15, 89-95.

Chapter 7

Trace Metal Content of Commercial Fertilizers Marketed in Lebanon

Isam Bashour, Ghada Hannoush, and Nasri Kawar

Faculty of Agriculture and Food Sciences, American University of Beirut, Beirut, Lebanon

> Recently, concern has been expressed over trace metals that enter the food chain. Inorganic fertilizers are considered among the potential sources of contamination. The majority of fertilizers marketed in Lebanon are imported from Europe and Middle Eastern countries. A total of 67 fertilizers samples were analyzed for Cd, Co, Cr, Ni, and Pb. Results show that most fertilizers contained low concentrations of these trace metals, but there is considerable variation in metal concentration detected among the samples. Highest concentrations of trace metals were found in granular phosphate sources followed by liquid formulations. Lowest concentrations of metals were found in soluble crystalline fertilizers. In general, the metal concentrations for Cd, Co, Cr. Ni and Pb were in the low in comparison to the range of values reported internationally for fertilizers.

Introduction

Inorganic fertilizers are used at high rates by Lebanese farmers in order to increase crop yields. Currently there are no regulations for fertilizer application rates or maximum concentrations of trace metals in fertilizers marketed in Lebanon. The concern of the public is increasing regarding the possibility of soil, water and food contamination with trace metals from excessive and repeated use of commercial fertilizers. Schroeder and Balassa (1963) noted that repeated fertilizer applications may raise the concentrations of some trace metals in food such as Cd that can be taken readily by plants and enter the food chain.

Allay (1971) and Kpomblekou and Tabatabai (1994) reported that the main source of fertilizer-derived trace metals in soils is phosphatic fertilizers manufactured from rock phosphate deposits. Soils naturally contain quantities of trace metals, due to weathering of the underlying parent material but additional accumulation may occur from sources like mineral fertilizers, manure and industrial activities. Many trace metals are not essential for plant growth, but can be absorbed by plants, to pass into the food chain and may cause health problems whenever they are present in high concentrations Oliver (1997).

Swaine (1962) compiled a comprehensive report about trace element content of fertilizers in the world and showed that all fertilizers contain trace metals but in varying amounts. Charter *et al.*, (1993) showed that P fertilizers such as triple super phosphate, diammonium phosphate and monoammonium phosphate contained variable concentrations of many trace metals.

Fertilizers in the Lebanese market are about 25% produced locally and about 75% imported from Europe and Middle Eastern Countries. No information is available about the concentration of trace metals in fertilizers marketed in Lebanon. This work was initiated to assess the content of specific trace metals Co, Cr, Cd, Ni and Pb in most of the fertilizer materials marketed in Lebanon. With this information it will be possible to begin to evaluate the risk of potential accumulation of these metals in Lebanese soils.

Materials and Methods

Between 1998 and 2000, 67 fertilizer samples were collected from fertilizers marketed in Lebanon. Samples were divided into 4 groups according to their nutrient content: N-fertilizers (12 samples); P- fertilizers (21 samples), K-fertilizers (9 samples) and NPK blends (25 samples). The samples were analyzed for their contents of Cd, Co, Cr, Ni and Pb according to the methods of analysis of the Association of Official Analytical Chemists (Williams, 2000).

A sample of 1.0g was dissolved in 5 mL concentrated HCl in a 25 mL beaker covered with watch glass, boiled for 10 minutes and then evaporated to near dryness. After cooling, the contents were boiled with 10 ml 0.1M HCl and quantitatively transferred into a 25 mL volumetric flask after filtration through Whatman no. 42 filter paper. The filtrate was analyzed for Cd, Cr, Co, Ni and Pb using GBC 902 atomic absorption spectrophotometer.

Results and Discussion

Concentrations of Cd, Cr, Co, Ni and Pb in the samples are presented in Tables I-III. The data indicated that concentrations of trace metals are highest in granular phosphatic fertilizers (SSP, TSP, DAP and MAP) followed by liquid phosphatic fertilizers. The water-soluble powder fertilizers, N-fertilizers and K-fertilizers generally contain relatively low concentrations of trace metals.

N-Fertilizers

The major sources of N fertilizers in the Lebanese market are ammonium sulfate (21%N), ammonium nitrate (33.5%N) and urea (46%N). Results of the analyses are presented in Table I & Figure 1 and show that the three N sources are relatively free of Cr and Co. Ammonium sulfate and ammonium nitrate did contain measurable amounts of Cd, Ni and Pb ranging between 1.5 – 7.2 mg Cd Kg^{-1}, <0.8 mg Ni Kg^{-1} and 2.2 – 3.0 mg Pb Kg^{-1}. These results were expected and were in agreement with the reported data of Al-Modaihesh and Al-Swailem (1998). Urea is a synthetic organic source of nitrogen, and therefore, it is not expected to contain trace metals. The sources of low concentrations of trace metals in NH_4NO_3 and $(NH_4)_2SO_4$ mostly due to the impurities in the nitric and sulfuric acids used in the production of these two salts. The data strongly suggest that the contents of trace metals in synthetic nitrogen fertilizers marketed in Lebanon will present little concern regarding environmental contamination from trace metal loading to soils, especially in the case of urea.

P- Fertilizers

The major sources of P fertilizers in the Lebanese market are: (1) locally produced granular single superphostatic, SSP (17% P_2O_5) and triple super phosphate, TSP (50% P_2O_5) and imported Jordanian diammonium phosphate, DAP (18–46–0); (2) Liquid phosphoric acid (52% P_2O_5) produced locally from Syrian rock phosphate and imported acidic liquid fertilizer (2–52–8) produced from green Jordanian phosphoric acid; and (3) soluble powder monoammonium phosphate, MAP (12–61–0) and diammonium phosphate, DAP (20-51-0) from Europe and South Africa. Liquids and soluble powders are commonly applied through irrigation systems (fertigation), while granular fertilizers are usually applied directly to the field. Trace metals concentrations in the phosphatic fertilizer samples are presented in Table II and figure 2. The data show that the

Table I. Trace concentrations (mg Kg^{-1}) in N-fertilizers

Sample No.	Description	Cd	Co	Cr	Ni	Pb
1.	Ammonium Sulfate	4.5	<0.1	<0.1	0.8	2.6
2.	Ammonium Sulfate	7.2	<0.1	<0.1	0.1	3.0
3.	Ammonium Sulfate	3.8	<0.1	<0.1	0.3	4.1
4.	Ammonium Sulfate	8.6	<0.1	<0.1	0.4	2.1
	Mean	**6.0**	**<0.1**	**<0.1**	**0.4**	**3.0**
5.	Ammonium Nitrate	2.3	<0.1	<0.1	0.2	2.9
6.	Ammonium Nitrate	3.1	<0.1	<0.1	0.1	2.0
7.	Ammonium Nitrate	1.8	<0.1	<0.1	0.1	1.8
8.	Ammonium Nitrate	1.5	<0.1	<0.1	0.3	2.5
	Mean	**2.2**	**<0.1**	**<0.1**	**0.2**	**2.3**
9.	Urea	<0.1	<0.1	<0.1	<0.1	<0.1
10.	Urea	<0.1	<0.1	<0.1	<0.1	<0.1
11.	Urea	<0.1	<0.1	<0.1	<0.1	<0.1
12.	Urea	<0.1	<0.1	<0.1	<0.1	<0.1
	Mean	**<0.1**	**<0.1**	**<0.1**	**<0.1**	**<0.1**

Note: The detection limit of the instrument for the analyzed trace elements is 0.01 mgL^{-1}. Therefore the detection limit values in the fertilizer samples are 0.1 mg Kg^{-1} because 1.0 g fertilizer sample was dissolved in 10 ml acid, filtered and directly measured using GBC 902 atomic absorption spectrophotometer.

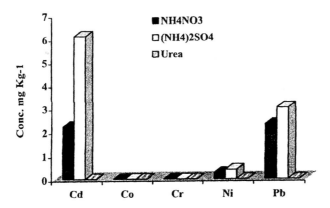

Figure 1. Means of trace metals in N-fertilizers marketed in Lebanon

concentrations of trace metals in liquid fertilizers are variable, probably because of variation in the trace metal content of the source H_3PO_4. The mean values in (mg Kg^{-1}) for liquid P sources were 19, 8.0, 56, 32 and 14 for Cd, Co, Cr, Ni and Pb respectively. The ranges of concentrations in mg Kg^{-1} were 1.1 – 31 for Cd, 1 – 13 for Co, 8.6 – 140 for Cr, 6.1 – 81 for Ni and 3.8 – 21 for Pb. These values were little lower than the reported values of Al-Modaihish and Al-Sewailem (1998) who analyzed fertilizer samples marketed in Saudi Arabia.

In manufacturing of single superphosphate, most of the trace metals contained in the original phosphate rock are usually found in the final fertilizer product. However, in manufacturing high analysis P-fertilizers, a great deal of the trace metals in the original phosphate rock are found in the by-product, gypsum. Rayment *et al.,* (1989) compared a small range of low and high analysis fertilizers manufactured in Queensland, Australia and found similar Cd:P ratio (413 ± 40 mg Cd Kg^{-1} P) in both high and low analysis formulations.

Analysis (Table II & Figure 2) indicated that granular P fertilizers contained higher concentrations of Cd, Cr, Ni, and Pb than liquids. The concentrations of Co in granular fertilizer were lower than concentrations of Co in the liquids. Mean values in mg Kg^{-1} for granular P sources were 30, 6.7, 79, 56 and 15 for Cd, Co, Cr, Ni and Pb, respectively. Ranges of concentrations were 21 – 41 for Cd, 2.8 – 10 for Co, 58 – 100 for Cr, 38 – 85 for Ni and 9.6 – 28 for Pb.

Concentrations of trace metals in water-soluble powder fertilizers were relatively low, below detection limits (<0.1 mg/kg). This could be attributed to the refined raw materials that were used in the production of these fertilizers. The results (Table II) indicated that mean values in mg Kg^{-1} for soluble powder fertilizers were 2.9 for Cd, 1.3 for Co, <0.1 for Cr, 2.2 for Ni and 0.9 for Pb. The main source of trace metals in liquid phosphatic fertilizers is H_3PO_4.

Table II. Trace metal concentrations (mg Kg^{-1}) in P- fertilizers

Sample No.	Formulation		Cd	Co	Cr	Ni	Pb
1.	18-46-0	Granular	28	5.8	88	52	10
2.	18-46-0	Granular	21	3.2	75	46	12
3.	18-46-0	Granular	22	2.8	70	38	11
4.	10-50-0	Granular	30	8.5	60	44	11
5.	11-52-0	Granular	25	7.8	58	40	9.6
6.	0-46-0	Granular	35	8.2	78	55	17
7.	0-46-0	Granular	33	5.6	86	61	16
8.	0-17-0	Granular	41	10	100	80	24
9.	0-17-0	Granular	39	8.4	98	85	28
		Mean	**30**	**6.7**	**79**	**56**	**16**
10.	H$_3$PO$_4$	Liquid	12	1.0	140	81	7.5
11.	H$_3$PO4	Liquid	31	12	50	6.1	15
12.	H$_3$PO4	Liquid	1.1	1.0	8.6	7.1	3.8
13.	H$_3$PO4	Liquid	3.5	4.1	25	5.3	·12
14.	2-52-8	Liquid	27	13	65	40	22
15.	2-52-8	Liquid	22	11	50	38	19
16.	8-48-0	Liquid	25	10	55	39	20
17.	12-46-0	Liquid	30	12	60	37	16
		Mean	**19**	**8.0**	**56**	**32**	**14**
18.	20-50-0	Powder	2.5	3.1	<0.1	2.0	1.2
19.	20-50-0	Powder	3.1	3.0	<0.1	3.2	1.8
20.	12-60-0	Powder	1.8	2.5	<0.1	3.2	1.8
21.	12-60-0	Powder	4.1	2.8	<0.1	2.8	0.5
		Mean	**2.9**	**2.9**	**<0.1**	**2.8**	**1.3**

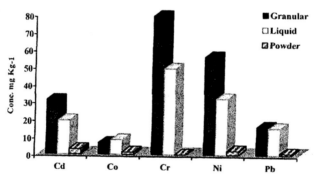

Figure 2. Means of trace metals in P-fertilizers marketed in Lebanon

K-Fertilizers

Potassium sulfate (50% K$_2$O) is the major K source in Lebanon, KCl (0-0-60) is not allowed in Lebanon due to high chloride levels. However, potassium nitrate (13-0-45), and to a lesser extent monopotassium phosphate (0-52-34) are used in drip irrigation, mainly on vegetables. Our analysis of trace metals in K-fertilizers in Lebanon is reported in Table III. All the K-fertilizers contained no detectable Cr and are low in Cd, Co, Ni and Pb content. Results of analysis (Table III) indicated that mean values in mg Kg^{-1} for granular potassium sulfate were 3.4 for Cd, 22 for Co, <0.1 for Cr, 7.1 for Ni and 10 for Pb.

Table III. Trace metal concentration (mg Kg⁻¹) in K-fertilizers

Sample No.	Formulation		Cd	Co	Cr	Ni	Pb
1.	0-0-50	Granular	3.5	23	<0.1	7.2	12
2.	0-0-50	Granular	2.8	19	<0.1	6.1	11
3.	0-0-50	Granular	4.0	25	<0.1	8.1	7.6
		Mean	3.4	22	<0.1	7.1	·10
4.	0-0-50	Powder	12	17	<0.1	2.3	0.5
5.	0-0-50	Powder	11	11	<0.1	4.8	3.8
6.	0-0-50	Powder	10	15	<0.1	6.1	8.1
7.	13-0-45	Powder	8.9	8.9	<0.1	1.1	8.9
8.	13-0-45	Powder	9.6	11	<0.1	1.1	8.9
9.	13-0-45	Powder	6.0	5.9	<0.1	4.7	4.4
		Mean	9.6	12	<0.1	3.4	5.8

NPK- Fertilizers

NPK compound fertilizers in Lebanon are marketed in granular, liquid and soluble forms. Trace metal concentrations in granular NPK fertilizers were higher than their concentrations in liquids, and were the lowest in soluble powders (Table IV & Figure 3). Based on the results presented in Tables II & IV it would appear that P-containing material is the dominant source of trace metals found in the NPK fertilizers.

Table IV. Trace metal concentration (mg Kg⁻¹) in NPKs

No.	Description		Cd	Co	Cr	Ni	Pb
1.	17-17-17	Granular	18	11	83	48	25
2.	17-17-17	Granular	20	15	80	33	20
3.	17-17-17	Granular	22	20	60	30	15
4.	18-18-18	Granular	15	5.5	89	39	44
5.	18-18-18	Granular	28	14	107	59	77
6.	15-15-15	Granular	19	9.8	77	33	28
7.	15-15-15	Granular	18	12	76	29	20
8.	15-15-15	Granular	31	10	60	25	18
9.	16-40-6	Granular	33	7.0	123	33	9.9
10.	30-10-10	Granular	5.5	3.0	40	13	4.8
		Mean	21	11	80	34	26
11.	25-25-18	Liquid	15	9.9	33	30	16
12.	25-25-18	Liquid	12	6.7	38	23	11
13.	24-24-18	Liquid	14	8.0	31	18	7.9
14.	7-5-5	Liquid	4.2	2.4	1.1	3.0	2.4
15.	6-5-4	Liquid	3.4	2.1	0.5	2.0	1.0
16.	2-8-28	Liquid	11	7.5	38	9.8	5.5
		Mean	10	6.1	22	14	7.4
17.	20-20-20	Powder	11	12	1.2	1.2	9.9
18.	20-20-20	Powder	13	11	2.1	3.4	7.0
19.	15-30-15	Powder	3.0	2.8	1.0	5.0	3.0
20.	15-30-15	Powder	14	9.6	4.0	6.6	10
21.	15-15-30	Powder	12	11	3.3	7.7	8.8
22.	22-7-7	Powder	4.2	5.5	6.0	2.1	3.2
23.	45630	Powder	5.0	1.1	4.2	2.0	2.0
24.	13-3-43	Powder	7.0	7.9	<0.1	<0.1	1.8
25.	28-14-14	Powder	12	5.0	0.8	<0.1	6.8
		Mean	9.1	7.3	2.5	3.1	5.8

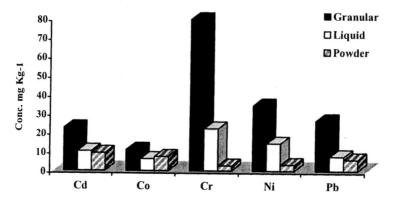

Figure 3. Means of trace metals in NPK-fertilizers

The main source of Cd added to agricultural soils is through application of phosphate fertilizers. Williams and David (1976) reported trace metals in P-fertilizers may accumulate in soil and become available for plant uptake. McLaughlin et al., (1996) reported that Cd accumulates in soil from applied fertilizers at a faster rate than Pb. Higher phytoavailability of Cd in soil alerted many researchers to this problem; therefore special attention was given to Cd in environmental research. The analysis of P-fertilizer samples in Table II indicate that Cd was present in granular phosphatic fertilizers 5.5 – 41 mg Kg^{-1}, and 1.1 – 21 mg Kg^{-1} in liquids. These values are similar to those reported in Wisconsin 1.5 – 9.7 mg Kg^{-1}; Iowa 6.8 – 47 mg Kg^{-1} (Lee and Keeney, 1975); 18-91 mg Kg^{-1} in Australia (Williams and David, 1973); > 0.1 –30mg Kg^{-1} in Sweden (Stenstromn and Vahter, 1974) and < 1 – 36.8 mg Kg^{-1} in Saudi Arabia with a median of 33.2 mg Kg^{-1} for granular P-fertilizers, 9.5 mg Kg^{-1} for liquids and 19.7 mg Kg^{-1} for granular NPK fertilizers (Al-Modaihish and Al-Sewailem, 1999).

The average annual application rates of fertilizers per hectare used on irrigated land in Lebanon are 700 Kg N-fertilizers, 500 Kg P-fertilizers and 150 Kg K-fertilizers (Bashour, 2000). The average Cd concentrations in fertilizers marketed in Lebanon is 2.7 mg Kg^{-1} in N-fertilizers, 21 mg Kg^{-1} in granular P-fertilizers and 7.6 mg Kg^{-1} in K-fertilizers (Table V & Figure 4). Therefore, about (700 x 2.7) + (500 x 21) + (7.6 x 150) = 13.51 g Cd ha^{-1} Yr^{-1} is typically applied to the Lebanese agricultural soils from fertilization with inorganic commercial fertilizers. Assuming that all applied Cd stays in the topsoil layer 0-30cm soil mass = 4400 tons ha^{-1}, then the increase in Cd concentration in the topsoil layer would be about 3 µg Kg^{-1}. This value is much lower than the tolerance limit in soils 0.2 mg Kg^{-1} in Germany, Al-Modaihish and Al-Sewailem 1998; 40 µg Cd Kg^{-1} in Washington State and Canada, Pan et. al. 2001.

Repeating the same calculations for Co, Cr, Ni and Pb using the means reported in Table V. the annual increase in Co, Cr, Ni and Pb from fertilizer application in Lebanon will correspond to 0.1, 6, 4.1 and 0.2 µg Kg^{-1} (Table VI).

Knowing that the soils in Lebanon are generally alkaline and calcareous with high clay content, trace elements may coprecipitate with carbonates as part of their structure or may be sorbed by oxides (mainly Fe & Mn) that precipitate onto carbonates or other soil particles. The greatest affinity for reaction with carbonates has been observed for Co, Cd, Cu, Fe, Mn, Ni, Pb, Sr, U and Zn (Kabata-Pendias, 2000). Therefore, it is expected that the free $CaCO_3$ in Lebanese soils will work as a trace element sink and carbonate compounds such as $CdCO_3$ is likely to occur if these soils are polluted with Cd. The present data Tables II & IV indicate that the measured trace elements are generally much lower than the tolerance limits in many developed countries such as Germany, Washing State and Canada (Table VI). Thus it indicates that, there is no danger from the contamination of Lebanese soils with trace metals from fertilization, provided the same quality of fertilizers will be marketed in Lebanon. Other possible sources of heavy metals are organic manure, sewage sludge, and aerial deposition. If a different quality of P-fertilizers were to be introduced to the market, environmental pollution from application of fertilizers will change.

The rock phosphate used for the production of SSP and TSP in Lebanon is imported from Syria and contains low Cd levels; 5 mg Cd Kg^{-1} P, the Cd/P ratios are among the lowest in the world (36–38) (Mc Laughlin et al., 1996). This explains the low Cd values in granular P-fertilizers produced in Lebanon.

CONCLUSION

From the analyses of 67 fertilizer samples marketed in Lebanon, it could be concluded that risks of food chain contamination by Cd, Co, Cr, Ni and Pb in commercial inorganic fertilizers are low because the concentrations of these trace metals are much lower than acceptable limits in many countries (Table VI).

International regulations for concentrations of metal impurities in fertilizers relates predominantly to Cd and sometimes to Pb. Limits of Cd in (mg Cd Kg^{-1} P) in P-fertilizer range from 0 in England to 450 mg Cd Kg^{-1} P in Australia, while the limits for Pb are 500mg Kg^{-1} (product basis) for all fertilizers and 10 mg Kg^{-1} for soil amendments (McLaughlin *et al.*, 1996).

Although the concentrations of heavy metals in fertilizers marketed in Lebanon are lower than the international limits, it is imperative that Lebanon should legislate for trace metals contents of fertilizers so as to safeguard the environment. Measurement of long-term accumulation of trace metals from organic and inorganic sources in agricultural land is also important so as to determine the acceptable loading limits of trace metals under Lebanese agricultural conditions.

References

1. Allaway, C.H. Feed and Food quality in relation to fertilizer use, In: R.A. Olson (ed) *Fertilizer Technology and Use*. Soil Sci. Soc. Amer. Madison, WI, 1971, pp 533-566.

Table V. Ranges and means of trace metal concentrations in mg Kg^{-1} for 67 fertilizer samples marketed in Lebanon

Product		Cd	Co	Cr	Ni	Pb
N-Fertilizers	Range	0-8.6	<0.1	<0.1	<0.1-0.8	<0.1-4.1
	Mean	2.7	<0.1	<0.1	2.0	1.7
P-Fertilizers	Range	1.1-41	1.0-13	0-140	2.0-85	0.5-28
	Mean	21	6.6	55	36	12
K-Fertilizers	Range	2.8-12	5.9-25	<0.1	1.1-8.1	0.5-12
	Mean	7.6	15	<0.1	4.6	2.3
NPK-Fertilizers	Range	3.4-33	1.1-20	<0.1-123	<0.1-59	1.0-77
	Mean	14	8.5	38	18	14

Table VI. Comparing possible trace metal cumulative additions from commercial fertilizers marketed in Lebanon with tolerance limits of Washington State and Canada

Element	Possible Cumulative metal addition to soil from commercial fertilizers application in Lebanon			Maximum acceptable cumulative metal addition to soil*	
	($\mu g\ kg^{-1}$)	($g\ ha^{-1}\ yr^{-1}$)	($kg\ ha^{-1}\ 45yrs^{-1}$)	Washington State ($g\ ha^{-1}\ yr^{-1}$)	Canada ($kg\ ha^{-1}\ 45\ yr^{-1}$)
Cd	4.0	18.5	0.84	89	4
Co	0.1	0.47	0.02	667	30
Cr	6.0	27.8	1.26	-	-
Ni	4.1	19.0	0.86	800	36
Pb	0.2	0.94	0.04	2222	100

Pan et al., 2001.

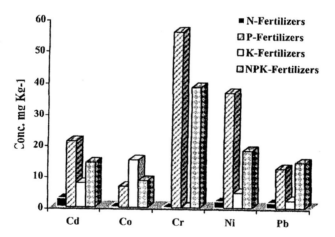

Figure 4. Means of trace metal concentrations in 67 fertilizer samples marketed in Lebanon

2. Al-Modaihish, A.S.; Al-Sewailem, M.S. Heavy metals content of commercial inorganic fertilizers used in the Kingdom of Saudi Arabia. Proceedings, 4th International Conference on Precision Agriculture 19-22 July, University of Minnesota, Min, USA, 1998, pp 1745-1754.
3. Bashour, I. Plant nutrient management for intensification of food production in the Near East Region. Agriculture, Land and Water Use Commission for the Near East, FAO Regional Conference March 25 – 27, 2000, Beirut, Lebanon, 2000, pp 1-15.
4. Charter, R.A.; Tabatabai, M.A.; Schafer, J.W. Metal content of fertilizer marketed in Iowa. *Comm. Soil Sci. Plant Anal*, 1993, 24:961-972.
5. Kabata-Pendias, Alina, Trace elements in soils and plants, 3rd ed., CRC press, LLC, N.W. Corporated Blvd., Boca Raton, Florida 33431, 2000.
6. Kpomblekou, A.K.; Tabatabai, M.A. Metal contents of phosphate rocks. *Comm. Soil Sci. Plant Anal.* 1994, 25: 2871-2882
7. Lee, K.W.; Keeney, D.R. Cadmium and zinc additions to Wisconsin soils by commercial fertilizers and west sludge application. *Water Air and Soil Pollution*, 1975, 5:109-112.
8. McLaughlin M.J.; Tiller, K.G.; Nacdue, R.; Stevens, D.P. Review: The behaviour and environmental impact of contaminants in fertilizers. *Aust. J. Soil Res.*, 1996, 34: 1- 54.
9. Oliver, M.A. Soil and human health: A review. *European Journal of Soil Sci.*, 1997, 48:573-592.
10. Pan, W.L., R.G. Stevens and K.A. Labao. 2001. Cadmium and Lead uptake by wheat and potato from P and waste-derived Zn fertilizers. 222nd American Chemical Society meeting Aug. 26-31, 2001, Chicago, USA.
11. Rayment G.E.; Best E.K.; Hamilton, D.J. Cadmium in fertilizers and soil amendments. Proc. RACI Chemistry International (1st Environmental Chemistry Division Conference), Brisbane, August 1989.
12. Schroeder, H.A.; Balassa, J.J. Cadmium: uptake by vegetables from super phosphate and soil. *Science*, 1963, 140: 819-820.
13. Stenstrom, T.; Vahter, M. Cadmium and lead in Swedish commercial fertilizers. *Ambio*, 1974, 3:90-91.
14. Swain, D.J. The trace–element content of fertilizers. Technical communication No. 52. Commonwealth Agricultural Bureaux, Farnham Royal, Bucks, England, 1962.
15. Williams, C. H.; David, D.J. The effect of superphosphate on the cadmium content of soils and plants. *Aust. J. Soil Res.*, 1973. 11: 43-56.
16. Williams, C.H.; David, D.J. The accumulation in soil of cadmium residues from phosphate fertilizers and their effect on the cadmium content of plants. *Soil Sci.*, 1976, 121: 86-93.
17. Williams, H. *Official Methods of Analysis of AOAC International.* (17th Ed). Chapter Editor Kane, P.F. Gaithersburg, Maryland, USA, 2000.

Chapter 8

Modeling the Distribution of Aluminum Speciation in Soil Water Equilibria with the Mineral Phase Jurbanite

C. Y. Wang[1,2], S. P. Bi[1,*], W. Tang[1], N. Gan[1], R. Xu[1], and L. X. Wen[1]

[1]Department of Chemistry, State Key Laboratory of Pollution Control and Resource Reuse of China, Nanjing University, 210093, Peoples Republic of China
[2]Department of Chemistry, Xuzhou Normal University, 221009, Peoples Republic of China
Corresponding author: email: bisp@nju.edu.cn

This paper presents the use of an equilibrium-based computer model to investigate the speciation of Al in soil solutions assumed to be in equilibrium with mineral phase jurbanite. The model predicts the distribution of various inorganic Al species, Al-organic matter complexes, and polymeric-Al species in solution as a function of pH. Using data from several published sources, the model demonstrates how change in soil solution composition impacts Al solution chemistry in equilbrium with several Al solid phases: jurbanite, basaluminite, alunite and gibbsite. Emphasis is placed on jurbanite due to its presence in soils impacted by acidic deposition. In the presence of jurbanite, the model predicts that SO_4^{2-} will have a substantial influence on the distribution of Al species and concentrations of soluble Al, while concentrations of organically complexed and fluoride complexed Al are minimal in the pH range studied. The model also predicts Al speciation for published soil solution data, assuming soil solutions were in equilibrium with jurbanite. Predicted concentrations of total dissolved Al, inorganic Al and Al-organic complexes agreed within an order of total dissolved Al, inorganic Al and Al-organic complexes agreed within an order of magnitude, however, the model consistently over predicts concentrations using the current set of constants. Nevertheless, the model results imply the presence and dissolution of jurbanite in soils impacted by acidic deposition will markedly influence soil solution Al chemistry.

Introduction

The release of toxic Al^{3+} has become one of the most serious consequences of anthropogenic soil acidification (*1*). Speciation of aluminum (Al) is a critical issue when assessing the effects of Al in soil solutions because not all chemical forms of Al are equally toxic. In order to better understand the effects of acid precipitation on soil and predict toxic concentration of Al in soil solutions, it is necessary to have a means to predict how the various forms of Al will respond to changes in the composition of soil solutions (*2*). Because of the widespread presence of solid phase gibbsite $Al(OH)_3$ in soils, control of Al solubility via gibbsite dissolution has been widely used in the modeling of soil solution chemistry (*3-5*). However, the Al concentration and the forms of Al species in soil solutions are related to the type of soil solid present, composition of the soil solution and soil pH (*6,7*). Significant changes in soil solution composition, such as a change in the dominant anion species, will impact Al solution chemistry and possibly the solid phase Al species present.

Acidic deposition represents one mechanism whereby atmospheric pollutants can influence soil chemistry, especially as H_2SO_4 is one of the most important components of acidic rainfall and acidic surface water (*8,9*). Introduction of acidic rainwater into the soil results in a significant change in soil solution composition, especially in the concentration of SO_4^{2-}. Similar changes occur for soil in contact with pyrite or amended with sulfur, as oxidation reactions result in the release of sulfate to the soil solution. With an increased concentration of sulfate in the soil solution, the activity of Al is greatly modified (*10-12*). The presence of sulfate may also change the relative stability of Al-containing minerals in the soil. In acid sulfate surface waters, aluminum oxysulfate minerals are more stable than gibbsite (*13*). At low pH values and higher SO_4^{2-} activities, jurbanite $Al(SO_4)(OH) \cdot 5H_2O$ appears to be the most stable phase of oxysulfate that will form in soils (*14-21*). The formation and presence of jurbanite has been reported in the B-horizon of soils(*13,15*). Models that predict Al speciation in soil solutions for soils impacted by acidic deposition, therefore, should take into account the possible presence of aluminum oxysulfate minerals such as jurbanite.

Direct measurement of the various potential Al species that may be present in soil solution is time-consuming and often not complete. Most often only a measure of the total dissolved Al concentration is possible. Computer-based chemical speciation models which assume chemical equilibrium in soil solutions are a simple and convenient way to predict the individual concentration of Al specie that may be present. In this paper we explore the use of a computer model to determination the distribution of Al species in soil solutions in equilibrium with the presence of solid phase jurbanite. The objectives of this effort are: to characterize the distribution of Al species in soil solutions; to evaluate the effects that changes (pH, concentrations of SO_4^{2-}, and solubility of jurbanite, etc.) in soil solution composition will have on Al speciation, and to compare predicted versus actual concentrations of total dissolved Al, inorganic Al and Al-organic complexes for published soil solution data.

Theory

Similar to our previous work (3,22-24), the model was constructed with the following assumptions:

(1) The concentration of Al^{3+} in soil water is controlled by the solubility of jurbanite:

$$Al(SO_4)(OH) \cdot 5H_2O + H^+ \rightleftharpoons Al^{3+} + SO_4^{2-} + 6H_2O \quad \log_{10}K_{sp} = -3.52$$

This reaction exists in a finite range of pH, when pH and C_{SO4}^* fit the condition that pH>$\log_{10}$ ($K_{s1}K_{sp}/C_{SO4}^*$) (K_{s1}, C_{SO4}^* see reference(3,23)). Setting $C_{SO4}^* = 1 \times 10^{-4}$ mol·L^{-1} (SO_4^{2-} concentration in natural waters range from 5.0×10^{-6} to 2.5×10^{-4} mol·L^{-1} based on our monitoring for fifty soil solution samples), jurbanite will be in its stable phase and doesn't dissolve at pH below 3.5.

(2) For the sake of simplicity, the studied soil solution is assumed as a dilute solution system with low ionic strength, so the effect of ionic strength need not be taken into account (3,25).

(3) Natural occuring organic acid in soil waters is depicted as a trinary acid proposed by Schecher and Driscoll (25). Two kinds of organically complexed Al (AlOrg, AlHOrg$^+$) are taken into consideration. In this paper we use 0.43 mol Org·mol^{-1} DOC (dissolved organic carbon) (3,25,26).

(4) Polynuclear Al is assumed to only exist in the form of dimer $Al_2(OH)_2^{4+}$ and trimer $Al_3(OH)_4^{5+}$ in acidic soil solutions. Al-phosphate complexes are expressed as $AlH_2PO_4^{2+}$(26) and the concentration of [$H_2PO_4^-$] is set at an extreme limitation value of highly concentration for forest soil of 1×10^{-4} mol·L^{-1}. Al-silicate complexes can be ignored (3).

(5) Al-silicate-complexes are ignored because of their very weak complexing properties and the lack of adequate equilibrium constants (3,26,27).

The mass balance in equilibrium with solid phase jurbanite list in Table I. The computer program is performed by using FORTRAN 77 language on a Pentium II computer. The input variable are total concentration of various ligands (C_F^*, C_{SO4}^* and C_{Org}^*), K_{sp} and pH. A Newton-Raphson iteration method was employed. The temperature is set as a constant 25°.

Table I. Mass balance in equilibrium with the presence of solid phase jurbanite

$C_{SO4}^* = [SO_4^{2-}] + [Al(SO_4)^+] + 2[Al(SO_4)_2^-]$

$C_F^* = [F^-] + [HF] + [AlF^{2+}] + 2[AlF_2^+] + 3[AlF_3] + 4[AlF_4^-] + 5[AlF_5^{2-}]$

$C_{Org}^* = [Org^{3-}] + [HOrg^{2-}] + [H_2Org^-] + [H_3Org] + [AlOrg] + [AlHOrg^+]$

C_{Al}^*
$= [Al^{3+}] + [Al\text{-}SO_4] + [Al\text{-}OH] + [Al\text{-}F] + [Al\text{-}Org] + [Al\text{-}Poly] + [Al\text{-}PO_4]$

in which:

$[Al^{3+}] = K_{sp}[H^+] / [SO_4^{2-}]$

$[Al\text{-}SO_4] = [Al(SO_4)^+] + [Al(SO_4)_2^-] = K_{sp}[H^+](K_{s1} + K_{s2}[SO_4^{2-}])$

$[Al\text{-}OH] = [Al(OH)^{2+}] + [Al(OH)_2^+] + [Al(OH)_4^-]$
$= K_{sp}[SO_4^{2-}]^{-1}(K_1 + K_2[H^+]^{-1} + K_4[H^+]^{-3})$

$[Al\text{-}F] = [AlF^{2+}] + [AlF_2^+] + [AlF_3] + [AlF_4^-] + [AlF_5^{2-}]$
$=$
$K_{sp}[H^+][SO_4^{2-}]^{-1}(K_{F1}[F^-] + K_{F2}[F^-]^2 + K_{F3}[F^-]^3 + K_{F4}[F^-]^4 + K_{F5}[F^-]^5)$

$[Al\text{-}Org] = [AlOrg] + [AlHOrg^+] = K_{sp}[H^+][SO_4^{2-}]^{-1} [Org^{3-}]$
$(K_{o1} + K_{o2}[H^+])$

$[Al\text{-}Poly] = 2[Al_2(OH)_2^{4+}] + 3[Al_3(OH)_4^{5+}]$
$= 2K_{p1}K_{sp}^2[SO_4^{2-}]^{-2} + 3K_{p2}K_{sp}^3[H^+]^{-1}[SO_4^{2-}]^{-3}$

$[Al\text{-}PO_4] = [AlH_2PO_4^{2+}] = K_{PO4}K_{sp}[H^+][SO_4^{2-}]^{-1}[H_2PO_4^-]$

NOTE: C_{SO4}^*: total concentration of sulfate; C_F^*: total concentration of fluoride; C_{Org}^*: total concentration of organic substances; C_{Al}^*: total concentration of aluminum. The general means of the K_1, K_2, K_4, $K_{F1\text{-}5}$, K_{s1}, K_{s2}, K_{p1}, K_{p2}, K_{PO4}, K_{o1}, K_{o2} are the chemical equilibrium constants (Log_{10} values, 25^0C) for the corresponding chemical reaction equations mentioned above. The are the same as those listed in literature 3 and 23.

Results and Discussion

The Distribution of Al Speciation for Soil Waters in Contact with Jurbanite

The influence of pH on Al speciation is illustrated in Figure 1. In the pH range of 3.5 to 5.0, the dominant Al species is free Al^{3+}. When pH increases from 5.0 to 7.0, the concentration of $Al(OH)^{2+}$ and $Al(OH)_2^+$ reach their maximum and are a fraction of the dissolved Al. Beyond pH 7.0, the concentration of OH^- is elevated, and $Al(OH)_4^-$ is the most abundant species, representing up to 99% of total Al. Fig. 1 illustrates that the relative concentration of toxic species of Al (Al^{3+}, $Al(OH)^{2+}$, $Al(OH)_2^+$) is high in the acidic pH ranges, while the less toxic species of Al-F, Al-Org constitute only a small fraction of the total concentration of Al. This suggests that soil waters in equilibrium with jurbanite could be potential toxicity to fine roots of susceptible plants(28).

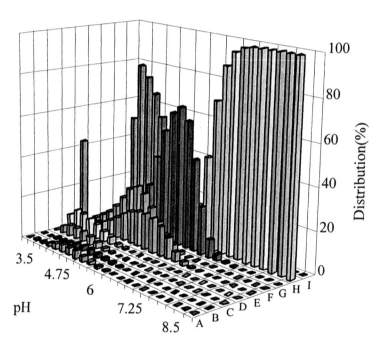

Figure 1. Distribution of various Al species as function of pH ($C_{SO4}^* = 1 \times 10^{-4}$ $mol \cdot L^{-1}$, $C_F^* = 5 \times 10^{-6}$ $mol \cdot L^{-1}$, $DOC = 1 \times 10^{-4}$ $mol \cdot L^{-1}$, $t = 25°C$). A: Al-F, B: Al-Org, C: Al-SO$_4$, D: Al-PO$_4$, E: Al-Poly, F: $Al(OH)^{2+}$, G: $Al(OH)_2^+$, H: $Al(OH)_4^-$, I: Al^{3+}.

Effect of SO_4^{2-} Concentration on the Distribution of Al Speciation and Dissolved Total Al Concentration

Our simulation shows that many factors, such as the concentration of SO_4^{2-}, F^- and DOC, may effectively affect the distribution of Al speciation in soil solutions. But F^- and DOC exert little influence on Al speciation when jurbanite is assumed present, whereas SO_4^{2-} takes on the dominant role in regulating Al speciation. As shown in Figure 2, when the concentration of SO_4^{2-} is decreased, the concentrations of toxic forms of free Al^{3+}, $AlOH^{2+}$ and $Al(OH)_2^+$ are greatly increased while the concentrations of Al-F, Al-SO_4, and Al-Org are reduced. This is because $[Al^{3+}] = K_{sp}[H^+] / [SO_4^{2-}]$, according to the solubility equilibrium reaction of jurbanite. Lower SO_4^{2-} concentration will promote the dissolution of jurbanite, resulting in an increase of in total soluble Al. This implies that the toxicity of Al in the soil waters in the presence of jurbanite will be significant as noted by Nilsson and Bergkvist (20). The formation of jurbanite in soil can thus be viewed as an intermediate step in the long chain of proton buffering processes causing a severe Al problem upon their dissolution (20).

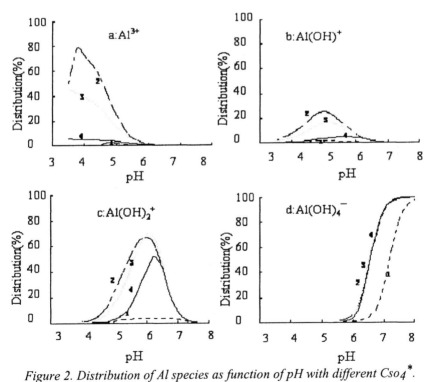

Figure 2. Distribution of Al species as function of pH with different C_{SO4}^*. ($C_F^* = 5 \times 10^{-6}$ mol·L^{-1}, DOC=1×10^{-4} mol·L^{-1}, $C_{PO4}^* = 1 \times 10^{-4}$ mol·L^{-1}, $t=25$ °C)
C_{SO4}^*: 1, 1×10^{-5} mol·L^{-1}; 2, 1×10^{-4} mol·L^{-1}; 3, 1×10^{-3} mol·L^{-1}; 4, 1×10^{-2} mol·L^{-1}.

Continued on next page.

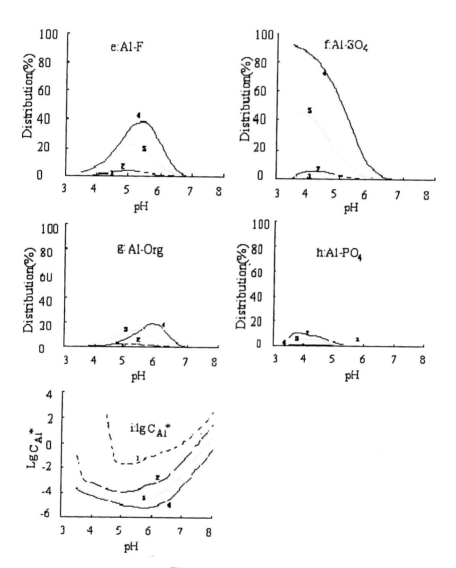

Figure 2. *Continued.*

Effect of Mineral Solubility on the Distribution of Al Speciation

The influence of changes in mineral dissolution on the distribution of Al species as a function of pH for jurbanite, basaluminite, alunite and gibbsite, have been reported in literature *(23)*. Basaluminite and alunite are oxysulfates that may or may not form in soil environments *(22-24)*. From our computer simulation at fixed sulfate concentrations, jurbanite will sustain a higher percentage of Al species in solution than the other oxysulfates or gibbsite for pH values<6.5. Above pH 6.5, there is little impact on mineral solubility on distribution of $Al(OH)_4^-$. While $Al-SO_4$ complexes were a significant fraction of the dissolved Al species in the presence of jurbanite (Fig. 2f), they were noticeably absent in computer simulations using basaluminite, alunite or gibbsite as the assumed controlling solid phase for the Al dissolution reaction. Polymeric Al species were also more evident in the presence of jurbanite for the pH range studied (pH 4-6). Calculated concentration of Organic-Al complexes varied substantially with mineral phase assumed controlling Al solubility. For the selected input concentration of ortho-phosphate, Al phosphate complexes were only significant at pH values < 4.75. Total dissolved Al concentration did show similar trends in concentration as a function of pH for basaluminite, alunite and gibbsite, with maximum concentrations above pH 4.3.

Application of Model to Soil Water Samples

The ability of the computer simulation to predict Al speciation in soil solutions was tested using published data for the composition of soil solutions extracted from acid soils *(2, 20)*. In all cases it was assumed that the measured Al in solution was in equilibrium with jurbanite. The results of the computer simulations predict that the dominant Al species in solutions are inorganic monomeric forms of Al (labile Al), while organically complexed Al (non-labile Al forms) was a minor component of the measured total dissolved Al concentration (Table II). The distribution of fluoride-complexed and sulfate-complexed Al varied between samples, reflecting differences in the composition of the original soil solutions samples, especially solution pH. As a point of reference, the calculated concentrations of monomeric Al^{3+}, $Al(OH)^{2+}$, and $Al(OH)_2^+$ are greater than the critical concentration of Al ($LC_{50} = 2 \mu mol \cdot L^{-1}$) reported for wheat seedlings *(29)*. These calculations support the assumption that in the presence of jurbanite, many soils could maintain concentrations of Al in soil solution which may be toxic to root growth when soil pH<5.0.

An assessment of the accuracy of the computer simulations was obtained by comparison of the calculated and measured concentrations of total dissolved inorganic Al (Al-Inorg), Al complexed with dissolved organic matter (Al-Org), and total dissolved Al (Total Al) (Table III). Predicted concentrations (B) of total dissolved Al, inorganic Al and organic complexes agreed within an order of magnitude with reported measured values (A), but the model consistently over predicts dissolved Al concentrations using the current set of equilibrium constants. The calculated percent difference between predicted and observed Al concentrations ((B-A)/A*100) was positive for 34 of the 42 pairs of observations in Table III, with the average percent difference for Al species ranging from +31 to +55%. This suggests that the solubility product for pure jurbanite may over predict the solubility of natural jurbanite formed in soils. Nevertheless, the model results are consistent with the hypothesis that the presence and subsequent dissolution of jurbanite in soils impacted by acidic deposition will markedly influence soil solution Al chemistry.

Table II. Calculated Al speciation for soil solutions of known composition (unit: $\mu mol \cdot L^{-1}$, except for pH)

No.	Al^{3+}	$Al(OH)^{2+}$	$Al(OH)_2^+$	Al-F	Al-SO$_4$	F^-	SO_4^{2-}	DOC	pH
1	10.9	1.31	0.18	24.4	4.41	167	464	10	4.5
2	5.68	3.42	2.31	23.7	0.86	63	177	10	5.2
3	92.6	3.52	0.15	10.1	13.7	27	185	10	4.0
4	63.2	3.81	0.26	14.3	8.61	33	167	10	4.2
5	31.5	3.01	0.32	30.8	5.45	89	206	10	4.4
6	23.1	5.54	1.49	12.9	2.16	19	111	10	4.8
7	9.57	3.64	1.55	25.4	1.37	61	167	10	5.0
8	22.7	2.16	0.23	26.5	5.48	107	284	10	4.4
9	50.3	4.80	0.51	21.6	5.42	38	131	10	4.4
10	24.2	3.66	0.62	21.3	3.43	49	168	10	4.6
11	73.5	17.90	4.19	4.98	15.6	8.1	220	8.1	4.4
12	78.7	23.50	6.69	4.98	12.7	6.2	169	6.2	4.5
13	67.1	18.60	4.92	4.97	13.7	13.7	211	13.7	4.4
14	68.0	23.50	7.76	4.97	11.0	4.4	167	4.4	4.5

NOTE: No.1~10 data from reference 2; No.11~14 data from reference 20.

Table III. Comparisons of predicted versus measured concentration of total dissolved Al (total Al), total dissolved inorganic Al (Al-Inorg), and Al complexed with dissolved organic matter (Al-Org)
(unit: µmol·L⁻¹, except for Percent Diff of unit: %)

No	Al-Org			Al-Inorg			Total Al		
	A	B	Percent Diff	A	B	Percent Diff	A	B	Percent Diff
1	1	1.9	90.00	58	41	-29.30	65	43	-33.85
2	1	3.1	210.00	24	37	54.17	30	40	33.33
3	4	2.5	-37.50	96	121	26.04	112	124	10.71
4	3	2.6	-13.33	66	91	37.88	89	93	4.49
5	9	2.6	-71.11	33	71	151.15	41	74	80.49
6	2	3.1	55.00	31	47	51.61	34	50	47.06
7	5	3	-40.00	19	42	121.05	28	45	60.71
8	2	2.3	15.00	37	57	54.05	75	59	-21.33
9	2	2.8	40.00	34	84	147.06	38	87	128.95
10	2	2.8	40.00	36	54	50.00	48	57	18.75
11	14	4.7	-66.43	90	120	33.33	116	144	24.14
12	13	19	46.15	92	133	44.57	106	152	43.40
13	23	42	82.61	92	113	22.83	115	155	34.78
14	7.4	14	89.19	87	121	39.08	95	136	43.16
	Mean=		31.40			54.82			33.91
	StdDev=		75.66			45.23			40.80

NOTE: A is actual soil solution data; B is computer model calculated results. No.1~10 data from reference 2; No.11~14 data from reference 20.

Acknowledgements

This project is financed by the National Natural Science Foundation of China (Key Project No. 49831005 & 29777013), Visiting Fellowship from State Education Ministration at the State Key Laboratory of Pollution Control & Resource Reuse of China, Nanjing University and 973 Key Fundamental Research Project of State Science & Technology Ministration of China (G1999011801), and Research Foundation of Xuzhou Normal University. We are deeply indebted to Prof. Guolian Ji, Tianren Yu for their encouragement and support.

References

1. Simonsson, M.; Berggren, D.; Gustafsson, J. P. Solubility of aluminum and silica in spodic horizons as affected by drying and freezing. *Soil Sci. Soc. Am. J.* **1999**, *63*, 1116-1123.
2. Fernandez-Sanjurjo, M. J.; Alvarez, E.; Garcia-Rodeia, E. Speciation and solubility control of aluminum in soil developed from slates of the river sor watershed (Galicia, NW Spain). *Water Air Soil Pollut.* **1998**, *103*, 35-53.
3. Bi, S. P. Investigation of the factors influencing aluminum speciation in natural water equilibrium with the mineral phase gibbsite. *Analyst.* **1995**, *120*, 2033-2039.
4. Driscoll, C. T.; Schecher, W. D. The chemistry of aluminum in the environment. *Environ. Geochem. Health.* **1990**, *12*, 28-50.
5. Arp, P. A.; Ouimet, R. Aluminum speciation in soil solutions: equilibrium calculations. *Water Air Soil Pollut.* **1986**, *31*, 359-366.
6. Godbold, D. L.; Jentschke, G. Aluminium accumulation in root cell walls coincides with inhibition of root growth but not with inhibition of magnesium uptake in Norway spruce. *Physiol. Plant.* **1998**, *102*, 553-560.
7. Tombacz, E.; Dobos, A.; Szekeres, M. Effect of pH and ionic strength on the interaction of humic acid with aluminum oxide. *Colloid and Polymer Sci.* **2000**, *278*(4), 337-345.
8. Prenzel, J. in *Effects of Accumulation of Air Pollutants in Forest Ecosystems*, D. Reidel Publ. Com. **1983**, pp. 157-170.
9. Merino, A.; Macias, F.; Garcia-Rodeja, E. Element fluxes and buffer reactions in acidified soils from a humid-temperate region of southern Europe. *Water Air Soil Pollut.* **2000**, *120*, 217-228.
10. Khanna, P. K.; Prenzel, J.; Meiwes, K. J.; Ulrich, B. Dynamics of sulfate retention by acid forest soils in an acidic deposition environment. *Soil Sci. Soc. Am. J.*, **1987**, *51*, 446-452.
11. Ulrich. B. In Hutchinson T. C.(eds.), *Effects of Acid Precipitation on Terrestrial Ecosystem.* Plenum Press, New York, 1980, p.255.
12. Nordstrom, D. K. The effect of sulfate on aluminium concentrations in natural waters: some stability relations in the system $Al_2O_2-SO_3-H_2O$ at 198K. *Geochimica Cosmochimica Acta.* **1982**, *46*, 681-692.
13. Eary, L. E. Geochemical and equilibrium trends in mine pit lakes. *Applied Geochemistry.* **1999**, *14*, 963-987.
14. Ludwig, B.; Prenzel, J.; Khanna, P. K. Modeling of cations in some acid soils from different acid input areas. *European J. Soil Science.* **1998**, *49*, 437-445.
15. Simonsson, M.; Berggren, D. Aluminium solubility related to secondary solid phases in upper B horizons with spodic characteristics. *European J. Soil Science.* **1998**, *49*, 317-326.

16. Levy, D. B.; Custis, K. H.; Casey, W. H.; Rock, P. A. The aqueous geochemistry of the abandoned spenceville copper pit, Nevada county, California. *J. Environ. Qual.* **1997**, *26*, 233-243.
17. Xu, S.; Harsh, J. B. Labile and non-labile silica in acid solutions: relation to the colloidal fraction. *Soil Sci. Soc. Am. J,* **1993**, *57,*1271-1277.
18. Shamshuddin, J.; Auxtero, E. A. Soil solution compositions and mineralogy of some active acid sulfate soils in Malaysia as affected by laboratory incubation with lime. *Soil Science.* **1991**, *152,* 365-376.
19. Courchesne, F.; Hendershot, W. H. The role of basic aluminum sulfate minerals in controlling sulfate retention in the mineral horizons of two spodosols. *Soil Science.* **1990**, *150,* 571-578.
20. Nilsson, S. I.; Bergkvist, B. B. Aluminium chemistry and acidification processes in a shallow podzol on the Swedish west coast. *Water Air Soil Pollut.* **1983**, *20*, 311-329.
21. Gundersen, P.; Beier, C. Aluminum sulfate solubility in acid forest soils in Denmark. *Water Air Soil Pollut.* **1988**, *39,* 247-262.
22. Bi, S. P.; An, S. Q.; Tang, W.; Xue, R.; Liu F. Computer simulation of the distribution of aluminum speciation in soil solutions in equilibrium with the mineral phase imogolite. *J. Inorg. Biochem.* **2001**, *87*, 97-104.
23. Bi, S. P.; An, S. Q.; Tang, W. Modeling the distribution of aluminum speciation in acid soil solution equilibrium with the mineral phase alunite. *Environmental Geology.* **2001**, *41*, 25-36.
24. Xue, R.; Duanmu, Y. J.; Wen, L. X.; Tang, W.; Bi, S. P. Computer simulation of aluminum speciation in soil water equilibria with mineral phase basaluminite. *Acta Scientiae Circumstantiae* (China). **2001**, *21*, 75-80.
25. Schecher, W. D.; Driscoll, C. T. An evaluation of the equilibrium calculations within acidification models: the effect of uncertainty in measured chemical components. *Water Resour. Res.* **1988**, *24*, 533-540.
26. Sposito G. *The Environmental Chemistry of Aluminum.* CRC Press; Inc. 2nd; Boca Raton, Florida, 1996, pp 87-116.
27. Westall, J. C.; Zachary, J. L.; Morel, F. M. M. *Technical Note 18*, Mass. Institute of Technology, Cambridge, 1976.
28. Comin, J. J.; Barloy, J.; Bourrie´, G.; Trolard, F. Differential effects of monomeric and polymeric aluminium on the root growth and on the biomass production of root and shoot of corn in solution culture. *European Journal of Agronomy.* **1999**, *11,* 115-122.
29. Wright, R. J.; Baligar, V. C.; Ahlrichs, J. L. The influence of extractable and soil solution aluminum on root growth of wheat seedlings. *Soil Sci.* **1989**, *148*, 293-302.

Chapter 9

Cadmium Accumulation in Wheat and Potato from Phosphate and Waste-Derived Zinc Fertilizers

W. L. Pan[1], R. G. Stevens[2], and K. A. Labno[1]

[1]Department of Crop and Soil Sciences, Washington State University, Pullman, WA 99164–6420
[2]Irrigated Agriculture Research and Extension Center, Washington State University, 24106 North Bunn Road, Prosser, WA 99350–9687

Phosphate and zinc fertilizer sources greatly vary in cadmium concentration, depending on the fertilizer raw material source and processing. Washington state has adopted Canadian guidelines for maximum allowable metal loading rates from fertilizers. Fertilizer screening rates are application rates established to determine potential metal loading rates in WA. Screening rates of 196 kg P_2O_5/ha and 8.4 kg Zn/ha were defined by 1998 Fertilizer Regulation Act. A 2-y field experiment was conducted to determine effects of some P and Zn fertilizers, applied at and above WA fertilizer screening rates, on wheat and potato Cd. An irrigated sandy soil was treated with 4 P sources, ranging from 49 to 780 mg Cd/kg P, and 1 waste derived Zn source. All sources applied at the WA screening rate maintained Cd levels at or below 0.05 mg Cd/kg$_{fw}$ for potato tubers and 0.1 mg Cd/kg$_{dw}$ for wheat grain. However, excessive triple super phosphate (TSP) applications over two years (1568 and 3156 kg P_2O_5/ha) exceeded 0.01 mg Cd/kg in wheat grain. In potato, 784 and 1568 kg P_2O_5/ha rates of TSP in both years and Western diammonium phosphate (WDAP) applied at 392 kg P_2O_5/ha over two years approached or exceeded 0.05 mg Cd/kg$_{fw}$ tuber. Overall, the current WA regulations on fertilizer Cd loading appear to be adequate at the established screening rates. Growers should be advised to adhere to agronomic rates to minimize metal loading.

Introduction

While there is a natural abundance of Cd in soils that is derived from the weathering of rocks and minerals, topsoils generally exhibit higher concentrations due to anthropogenic inputs such as industrial pollution, agronomic amendments in crop production (Holmgren et al., 1993; Kabata-Pendias and Adriano, 1995; McLaughlin et al, 1999) and plant recycling. Fertilizers are a significant source of Cd found in agricultural soils. Cadmium is a contaminant in many P and waste-derived micronutrient fertilizers and the Cd concentration varies greatly by the raw material source of the fertilizer (Mortvedt, 1985; Mortvedt, 1996). Globally, there are notable differences delineated by region of origin of the P ores. For example, western U.S. P deposits contain higher Cd levels compared to eastern U.S. sources (Kongshaug et al., 1992).

Several long-term field experiments have shown increases in soil and crop Cd accumulation over time with repeated inputs of Cd containing P fertilizers (Jones and Johnston, 1989; Andersson and Siman, 1991; Loganathan et al., 1995; Hamon et al., 1998; Gray et al., 1999). Long-term experiments in the U.S. have shown mixed results, with no evidence of increased crop Cd concentrations at some sites and modest increases at others (Mortvedt, 1987; Gavi et al., 1997), possibly attributable to a wide range of Cd contents in P fertilizers used in the U.S. In some comparisons, P treatments that increased yields may have also caused Cd dilution. Variations may also be attributed to soil chemical and biological factors that modify metal availability (; Kabata-Pendias and Adriano, 1995).

A fertilizer screening survey was conducted by the Washington State Department of Agriculture (Bowhay, 1997) to determine a range of trace metal concentrations contained in fertilizers typically used in WA state. In addition to several P fertilizers, waste-derived micronutrient fertilizers were identified as another source of Cd and other trace metals in agricultural systems. For example, two granular Zn fertilizers were estimated to contain 52 and 275 mg Cd/kg. Subsequently, $ZnSO_4$ (360,000 mg Zn/kg), imported from a fireworks manufacturer in China, was found to contain at least 120,000 mg Cd/kg (Cooper et al., 2001). Additional screening by the Idaho State Department of Agriculture (ISDA) revealed a $ZnSO_4$ stock with 66,000 mg Cd/kg. Field applications were made at rates that delivered up to 1.34 kg Cd/ha. This application rate far exceeds the WA state maximum allowable loading rate of 0.089 kg Cd/ha. The material has since been removed from the market.

The heightened public awareness of trace metals in inorganic fertilizers can be traced back to a series of events that occurred near Quincy, WA (Stevens, 1997; Wilson, 2001). Crop losses were reported between 1990 and 1996 on fields amended with wash water from farm chemical handling equipment and recycled waste. The losses were alleged by the affected producers to be related to trace metal contamination in the fertilizer rinsate, but herbicide residuals were suspected to cause the crop losses (Stevens, 1997). Nevertheless, public hearings and newspaper reports led to a broader public debate concerning the regulation of trace metal contaminants in processed fertilizers.

Deliberations by the Fertilizer Advisory Workgroup, established by Governor Gary Locke lead to the passage of the Fertilizer Regulation Act (Washington State Legislature, 1998), which set maximum acceptable annual metal additions from inorganic fertilizers, e.g. 0.089 kg Cd/ha/year. These limits were adopted from Canadian standards outlined in the 1996 Canadian Memorandum T-4-93, which established 45 year cumulative soil loading limits, and annual limits for Washington state were determined by dividing the Canadian cumulative limits by 45 years (Stevens, 1999). Washington screening application rates were then established to characterize current nutrient uses over 4-year crop rotations. For P and Zn fertilizers, 4-year cumulative screening application rates of 784 kg P_2O_5/ha and 33.6 kg Zn/ha are used to calculate metal loading rates. For a given fertilizer, annual screening rates (referred to as 1x rates, Table I) are used to determine the metal loading rates that would be co-applied with the plant nutrients. If the calculated loading rates are found to exceed the maximum acceptable annual metal addition rate, then additional labeling is required to limit application rates that meet metal loading rate standards.

The objectives of the present research were to provide an initial assessment of the newly adopted WA state regulations by determining the effects of selected P and Zn fertilizer sources applied at and above screening rates on Cd concentrations in wheat grain and potato tubers. These crops were chosen for this experiment since the U.S. FDA reported that approximately 40% of the dietary Cd intake of the "typical" American is consumed in the form of cereal grain and potato products (Chaney and Horneck, 1978).

Methods

A two-year randomized complete block field experiment was conducted on two adjacent sites, one site dominated by Quincy fine sand (mixed mesic Typic Torripsamment) and the other site dominated by Hezel loamy fine sand (mixed mesic Xeric Torriorthents) and Warden very fine sandy loam (mixed mesic Xeric Haplocambids) at the Washington State University Irrigated Agriculture Research Extension Center near Prosser, WA. The site has been annually cropped in small-grain, corn and alfalfa rotation for 40 years. Pre-treatment soil test values were: pH of 6.2, 0.83% organic matter, 5.9 meq/100g Na acetate extractable Ca, 10.5 mg bicarbonate-extractable P/kg and 1.0 mg DTPA extractable Zn/kg in the wheat plots and 15 mg P/kg and 2.0 mg Zn/kg in the potato plots. A pre-fertilization baseline level of 0.097 mg total soil Cd/kg in

Table I. Fertilizer sources, annual application rates of P and Zn, Cd concentrations and Cd rates.

Rate	Source	P_2O_5/Zn	Fertilizer	Cadmium
		(kg/ha/yr)	(mg/kg)	(g/ha/yr)
1x[a]	E DAP[b]	196	9.8	4.2
2x	E DAP	392	9.8	8.3
1x	W DAP	196	103	44
2x	W DAP	392	103	87
1x	RP	196	44	29
2x	RP	392	44	58
4x	RP	784	44	116
8x	RP	1568	44	232
1x	TSP[c]	196	150	65
2x	TSP	392	150	131
4x	TSP	784	150	262
8x	TSP	1568	150	524
1x	$ZnSO_4$[d]	8.4*	295	18
2x	$ZnSO_4$	16.8	295	32
4x	$ZnSO_4$	33.6	295	59
8x	$ZnSO_4$	67.2	295	114

[a] x = Screening rate for P or Zn.
[b] E DAP = Florida 18-46-0, W DAP = Idaho 18-46-0, RP = Idaho 30% rock phosphate,
[c] TSP = Idaho 0-45-0, $ZnSO_4$ = 18% granular zinc
[d] all $ZnSO_4$ treatments received 196 kg/ha/yr P_2O_5 and 4 g Cd/ha/yr from E DAP

the upper 15 cm was used to calculate total soil Cd in Figures 1 and 3. Phosphorus and Zn fertilizers were preplant broadcast-incorporated into 3.45 by 10.7 m plots to a depth of 15 cm with a rototiller. Fertilizer Cd application rates were calculated from fertilizer rates and Cd concentrations (Table I). The total Cd concentration of fertilizers was determined by digesting 1 g fertilizer in 20 ml aqua regia over night at room temperature, followed by Cd analysis by atomic absorption spectrophotometry.

Additional N fertilizer was spring top dressed in the first growing season and at the time of P and Zn fertilization prior to the second growing season to provide uniform total N application to all treatments.

Soil samples were taken from 0 to 15 cm previous to fertilization for soil test analysis by compositing 10 cores per replicate. Post-harvest soil samples were taken from individual plots at the 0 to 15 cm depth by compositing 5 samples per plot. Soil samples were air-dried and screened to pass a 2 mm sieve.

Soft white winter wheat (*Triticum aestivum* cv. Stephens) was planted in fall 1998 in plots dominated by Quincy fine sand, and managed with standard irrigated agronomic practices. Grain was machine combined for yield determination and subsampled for Cd analysis.

Potato (*Solanum tuberosum* cv. Russet Burbank) was planted in spring 1999 on plots adjacent to the wheat on Hezel-Warden series and managed with standard irrigated agronomic practices. Mature tubers were machine harvested and graded according to Washington State standards.

For the second growing season, fertilizer applications were repeated from the same sources and at the same rates on the same plots. Winter wheat, seeded in fall 1999 and grown to maturity in 2000 followed the 1999 potato crop, and potatoes were planted in spring 2000 following the 1999 wheat crop.

Harvest tubers were subsampled for analysis based on the weight ratio of the final tuber yield marketable category distribution so about 14 to 16 tubers were used. Tubers were scrubbed gently using soft nylon brushes and rinsed twice with tap water to remove soil particles. Tubers were cut length-wise and half of the tuber was sliced and dried for analysis. All plant sample material was oven dried at 60 C for at least 24 hours.

All glassware used for trace metal analysis was acid washed with 1:1 nitric solution. Plant and soil samples for total metal analysis were wet digested or extracted by soaking 1 g of tissue with 5 ml concentrated nitric acid (Baker Instra-analyzed reagent for trace analysis) overnight, followed by refluxing for 1 h at 120 C and three cycles of 3 ml 30% hydrogen peroxide addition and 30 min reflux (Jones and Case, 1990). Once cooled, digests were syringe filtered into plastic tubes. Trace metals were analyzed by inductively coupled argon plasma-mass spectrometry (Hewlett Packard Model No. 4500). Two method blanks and one National Institute for Standards and Technology (NIST) certified wheat sample (#8436) were included with every batch of 40 digested plant samples to assess metal recovery. Samples were corrected for mean method blank Cd concentration per run. Cadmium concentrations are reported on a dry weight basis for wheat grain and soil. Harvest potato tuber Cd concentrations are expressed on a fresh weight basis.

Results and Discussion
Wheat

Grain Cd concentrations ranged from 0.017 to 0.107 mg Cd/kg in 1999 after the first year of imposed fertilizer treatments on the Quincy fine sand. Only the 8x TSP application, which added 0.524 kg Cd/ha (Table I), or 6 times the maximum acceptable annual Cd addition exceeded 0.1 mg Cd/kg grain (Figure 1). Increasing rates of RP with corresponding increases in Cd input had no effect on grain Cd concentration, likely due to its relative insolubility.

In the second year of treatment imposition following the same treatments in potato the previous year, grain Cd was again related to total soil Cd, but the Cd concentrations increased overall in the second site-year. The grain Cd produced on the Hezel-Warden inclusion following the first year potato crop ranged from 0.074 to 0.198. This difference between site-years was associated with a similar difference in DTPA Cd between the site-years (Labno, 2001), and was apparent even for comparisons of treatments with equivalent total Cd input (e.g. 2x applied in year 1 vs. 1x applied in years 1 and 2). It is uncertain if this is attributable to environmental, cultural history or soil differences between the two years.

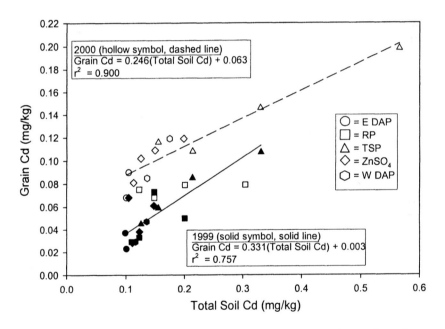

Figure 1. Relationship between wheat grain Cd and total soil Cd over all fertilizer treatments except RP. Total soil Cd was estimated as the summation of baseline total soil Cd at 0 to 15 cm and fertilizer Cd added over the first (1999) or both years (2000).

The range of grain Cd concentrations reported herein are similar to ranges reported for wheat produced in other regions of the world such as Britain (Chaudri et al., 1995, Sweden (Andersson and Petterson, 1981), Australia

(Oliver et al., 1995) and Canada (Grant and Bailey, 1998). Nevertheless, all 1x treatments in this second growing season except TSP sustained grain Cd levels below 0.1 mg Cd/kg. Transfer coefficients representing the change in grain Cd per unit change in total soil Cd were 0.331 in 1999 and 0.246 in 2000 over all treatments excluding RP, which had no effect on grain Cd. The TSP had the highest Cd/P ratio (0.75 g Cd/kg P) among all P fertilizers, resulting in grain Cd ranging from 0.107 to 0.198 mg Cd/kg grain in the second growing season. When averaged over the 1x and 2x rates applied in year 2, the TSP produced higher grain Cd (0.111 mg Cd/kg) than RP (0.071 mg Cd/kg) while the DAP sources did not significantly differ from one another or from the other P sources (Figure 2). This contrasts with other field comparisons of DAP fertilizers, where DAP with 153 mg Cd/kg increased wheat grain Cd from 0.028 to 0.086 mg Cd/kg over a low Cd DAP source with 2 mg Cd/kg (Mortvedt et al., 1981). The lack of differences in grain Cd concentrations from the DAP sources in the present study may have been due to overriding effects of soil chemical reactions affecting Cd availability at these lower levels of addition.

High rates of RP resulted in lower grain Cd than predicted from the overall relationship, reflecting lower RP solubility compared to other fertilizers in this system (Figure 1). Grain yield in the first growing season was significantly

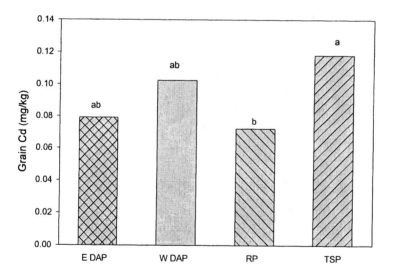

Figure 2. Grain Cd concentrations averaged over second year application of 1x and 2x rates of different P sources. Means with shared letters are not significantly different at Tukey's $HSD_{0.05}$.

Table II. Fresh weight grain and tuber yields after one year of fertilizer application (harvested 7/26/99 and 9/16/99 respectively at Prosser, WA).

Source	P_2O_5 or Zn	Grain	*	Tuber	
	—— (kg/ha) ——	—— (kg/ha) ——		—— (Mg/ha) ——	
EDAP	196	8800	a	74	NS
EDAP	392	8645	a	77.7	
RP	196	4305	e	76.8	
RP	398	4540	e	74.9	
RP	784	5458	cde	71.2	
RP	1568	5351	cde	71.6	
TSP	196	8414	a	72.18	
TSP	392	7956	ab	74.9	
TSP	784	9346	a	70.1	
TSP	1568	6944	abcd	65.2	
$ZnSO_4$	8.4	6763	bcde	78.1	
$ZnSO_4$	16.8	7582	abc	72.73	
$ZnSO_4$	33.6	8755	a	71.22	
$ZnSO_4$	67.2	8626	ab	70.5	
W DAP	196	8298	a	71.6	
W DAP	392	9057	a	66.7	

* shared letters not significantly different at Tukey's HSD 0.05 level of probability.
NS = not significant

lowered by the RP compared to other P sources (Table II), particularly the lower rates of RP. Visual symptoms of P deficiency, trends towards higher yields with increasing rates of RP, and a lack of yield depression with other P sources containing higher Cd suggest the yield depression was due to RP insolubility rather than Cd toxicity from this P source.

Applications of the $ZnSO_4$ up to the 8x screening rate for Zn in the first growing season did not raise grain Cd levels above 0.07 mg Cd/kg grain (Figure 1). This was partially attributable to the relatively low Cd loading rates with the Zn treatments, due to lower Zn application rates relative to P rates. Zinc fertilization of soil marginally to severely Zn deficient has been shown to decrease wheat grain Cd concentrations, implying the potential for a Zn-Cd antagonism (Oliver et al., 1994). However, the reapplication of the same rates for the second wheat crop following potatoes raised grain Cd to 0.074 for the 1x rate to 0.118 mg Cd/kg for the 8x rate. This represents a 16x rate of Zn application over two years, in which the cumulative Cd application rate approached those imposed by P treatments such as the 4 x TSP in the first year.

Potato

Unpeeled tuber Cd concentrations in the first growing season, expressed on a fresh weight basis, ranged from 0.023 mg Cd/kg tuber 0.052 mg Cd/kg tuber in the 4x TSP (Figure 3). For comparative purposes, Wolnik et al. (1983)

reported tuber Cd surveyed across the U.S. at 297 sites ranging from 0.002 to 0.182 mg Cd/kg with a mean of 0.031 mg Cd/kg. A similar range was reported for potatoes produced in Australia, ranging from 0.004 to 0.232 mg Cd/kg with a mean of 0.041 (McLaughlin et al., 1997). All 1x rates of applications sustained tuber Cd below 0.05 mg Cd/kg for both growing seasons.

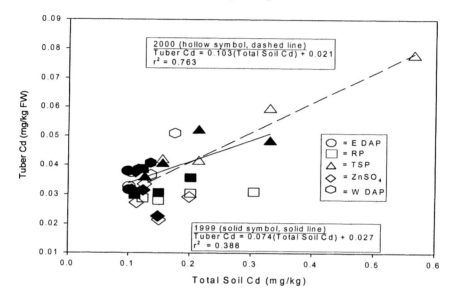

Figure 3. Relationship between potato tuber Cd and total soil Cd over all fertilizer treatments except RP. Total soil Cd was estimated as the summation of baseline total soil Cd at 0 to 15 cm and fertilizer Cd added over the first (1999) or both years (2000).

The only fertilizer applied at the 2x annual rate to exceed this value was the Idaho DAP applied in the second year, producing 0.052 mg Cd/kg. The P sources did not influence the tuber Cd when averaged over the 1x and 2x screening rates. In 2000, tuber Cd ranged from 0.022 to 0.078 mg Cd/kg. Excessive TSP rates of 4x and 8x re-applied before the second growing season elevated the tuber Cd to 0.059 and 0.078 mg Cd/kg. Similar to the wheat Cd relationships, RP tended to produce lower tuber Cd at a given soil total Cd level. In contrast, these P source differences were minimized in experiments conducted under more acidic conditions where RP solubility was likely higher (McLaughlin et al., 1995).

Transfer coefficients reflecting the increase in tuber Cd per unit increase in total soil Cd were similar for the two years: 0.074 for 1999 and 0.103 for 2000 (Figure 3). These coefficients are lower than those reported for Australian potato production, which ranged from 0.02 to 4.5 with a mean value of 0.5, based on EDTA-extractable Cd, which the authors claimed to be similar to total Cd for those soils (McLaughlin et al., 1997). Increasing Cd inputs with increasing rates of $ZnSO_4$ did not increase tuber Cd, and Cd concentrations did not exceed 0.05 mg Cd/kg, perhaps attributable to Zn-Cd antagonism in potato (McLaughlin et al., 1995). Total potato tuber yield averaged 72.7 Mg/ha (Table II) and 80.8 Mg/ha in the second year and was not significantly affected by fertilizer source or rate. Marketable categories were also not affected by any Zn or P treatment.

Summary and Conclusion

Cadmium concentration in wheat grain and potato tubers was affected by P fertilizer source and rate. Application of all fertilizers in this study at the WA screening rate maintained grain and tuber Cd levels at or below guide values for potato tubers (0.05 mg Cd/kg_{fw}) (McLaughlin et al., 1997) and wheat grain (0.1 mg Cd/kg_{dw}) (Norvell et al., 2000). However, application of the soluble P and Zn fertilizers, at 2x WA screening rate and above, attained or exceeded these guide values for wheat following potato. In potato, excessive rates of TSP (4x and 8x) in both years and the 2x rate of Western DAP in the second year approached or exceeded 0.05 mg Cd/kg_{fw} in potato tubers. Overall, the current WA regulations appear to be adequate for minimizing wheat and potato Cd concentrations with fertilizers applied at typical application rates. Growers should be advised to adhere to agronomic rates of fertilizer application to limit metal loading of agricultural soils.

References

1. Holmgren, G.G.S., Meyer, M.W., Chaney, R.L., Daniels, R.B. Cadmium, lead, zinc, copper, and nickel in agricultural soils of the United States of America, 1993, J.Environ.Qual., 22, 335-348.
2. Kabata-Pendias, A., Adriano, D. C. Trace metals, 1995, *In* Soil amendments and quality, J. Rechcigl (ed.) p. 139-167, Lewis Publ., Boca Raton, FL.
3. McLaughlin, M., Parker, D., Clarke, J. Metals and micronutrients – food safety issues, Field Crops Research, 1999, 60, 143-163.
4. Mortvedt, J. J. 1985. Plant uptake of heavy metals in zinc fertilizers made from industrial by-products. J. Environ. Qual. 14:424-427.

5. Mortvedt, J.J. Heavy metal contaminants in inorganic and organic fertilizers. Fert. Res. 1996, 43, 55-61.
6. Kongshaug, G., Bockman, O.C., Kaarstad, O., Morka, H. Input of trace elements to soils and plants, 1992, Olso. 21-22 May. Conf. of Chemical Climatology and Geomedical Problems. Norwegian Academy of Science and Letters, Norway, pp 185-216.
7. Jones, K.C., Johnston, A.E. Cadmium in cereal grain and herbage from long-term experimental plots at Rothamstedt, UK, Environ. Pollut., 1989, 57, 199-216.
8. Andersson, A., Siman. G. Levels of Cd and some other trace elements in soils and crops as influenced by lime and fertilizer level, 1991, Acta Agric. Scand. 41, 3-11.
9. Loganathan, P., Mackay, A.D. Lee, J., Hedley, M.J. Cadmium distribution in hill pastures as influenced by 20 years of phosphate application and sheep grazing, Aust. J. Soil Res., 33, 859-871.
10. Hamon, R.E., McLaughlin, M.J., Naidu, R., Correll, R. Long-term changes in cadmium bioavailability in soil, Env. Sci. Technol., 1998, 32, 3699-3703.
11. Gray, C.W., McLaren, R.G., Roberts, A.H.C., Condron, L.M. The effect of long-term phosphatic fertilizer applications on the amounts and forms of cadmium in soils under pasture in New Zealand, 1999, Nutrient Cycling in Agroecosystems, 1-11.
12. Mortvedt, J.J, Cadmium levels in soils and plants from some long-term soil fertility experiments in the United States of America, J. Environ. Qual., 1987, 16, 137-142.
13. Gavi, R., Basta, N., Raun, W. Wheat grain cadmium as affected by long-term fertilization and soil acidity, 1997, J. Environ. Qual., 26, 265-271.
14. Screening Survey for Metals in Fertilizers and Industrial By-Products Fertilizers in Washington State; Bowhay, D., Ed. Washington State Department of Ecology: Olympia, WA, 1997; Publication No. 97-342.
15. Cooper, M.E., Andereason, R., Baker, J.C. Idaho State Department of Agriculture Division of Plant Industries investigative report on cadmium contaminated zinc sulfate, 2001, Cadmium Research Network, Agrium, Calgary, Alberta.
16. Stevens, R. Fertilizers and Heavy metals in Washington state. Proceedings of the Western Nutrient Management Conference, 1999, 4, 98-105.
17. Wilson, D. Fateful Harvest, 2001, HarperCollins Publishers, New York, NY.
18. Chaney, R.L., Hornick, S.B. Accumulation and effects of cadmium on crops, 1978, In Proceedings of the First International Cadmium Conference, Cadmium Association (ed), Drogher Press for Metal Bulletins, England, 125-140.

19. Jones, J., Case, V. Sampling, handling, and analyzing plant tissue samples, 1990, In Westerman, R.L. (ed.) Soil Testing and Plant Analysis, SSSA, Madison, p. 389-427.
20. Labno, K.A., Cadmium and lead uptake by wheat and potato from phosphorus and waste-derived zinc fertilizers, 2001, M.S. Thesis, Washington State University, Pullman, WA.
21. Chaudri, A., Zhao, F.J., McGrath, S.P., Crosland, A.R. The cadmium content of British wheat grain, 1995, J. Environ. Qual., 24, 850-855.
22. Andersson, A., Pettersson, O. Cadmium in Swedish winter wheat, 1981, Swedish J.Agric. Res., 11, 49-55.
23. Oliver, D.P., Gartrell, J.W., Tiller, K.G., Correll, R., Cozens, G.D., Youngberg, B.L. Differential responses of Australian wheat cultivars to cadmium concentration in wheat grain, 1995, Aust. J. Agric. Res., 46, 873-886.
24. Grant, C.A., Bailey, L.D. Nitrogen, phosphorus and zinc management effects on grain yield and cadmium concentration in two cultivars of durum wheat, 1998, Can. J. Plant Sci., 74, 307-314.
25. Mortvedt, J.J., Mays, D.A., Osborn, G. Uptake by wheat of cadmium and other heavy metal contaminants in phosphate fertilizers, 1981, J. Environ. Qual., 10, 193-197.
26. Oliver, D.P., Hannam, R., Tiller, K.G., Wilhelm, N.S., Merry, R.H., Cozens, G.D. The effects of zinc fertilization on cadmium concentration in wheat grain, 1994, J. Environ. Qual., 23, 705-711.
27. Wolnik, K.A., Fricke, F.L., Capar, S.G., Braude, G.L., Meyer, M.W., Satzger, R.D., Bonnin, E., 1983, Elements in major raw agricultural crops in the United States. 1. Cadmium and lead in lettuce, peanuts, potatoes, soybeans, sweet corn and wheat, 1983, J. Agric. Food Chem., 31, 140-1244.
28. McLaughlin, M.J., Maier, N.A., Rayment, G.E., Sparrow, L.A., Berg, G., McKay, A., Milham, R.H, Merry, R.H., Smart, M.K. Cadmium in Australian potato tubers and soils, 1997, J. Environ. Qual., 26, 1644-1649.
29. McLaughlin, M., Maier, N.A., Freeman, K., Tiller, K.G., Williams, C.M.J., Smart, M.K. Effect of potassic and phosphatic fertilizer type, fertilizer Cd concentration and zinc rate on cadmium uptake by potatoes, 1995, Fert. Res. 40, 63-70.
30. Norvell, W.A., Wu, J., Hopkins, D.G., Welch, R.M. Association of cadmium in durum wheat grain with soil chloride and chelated extractable soil cadmium, 2000, Soil Sci. Soc. Am. J., 64, 2162-2168.

Chapter 10

Health Risk Assessment for Metals in Inorganic Fertilizers: Development and Use in Risk Management

Daniel M. Woltering[1]

The Weinberg Group Inc., 1220 19th Street NW, Washington, D.C. 20036
[1]Current address: Water Environment Research Foundation, 635 Slaters Lane, Suite 300, Alexandria, VA 22314

Chemical exposure and risk assessment is a widely used and accepted scientific practice. Its uses include product and environmental media evaluation as well as establishing safe limits for risk management. Responding to questions about the presence and safety of non-nutrient components in fertilizers, stakeholders have performed health risk assessments for NPK and micronutrient products. Information on fertilizer products and practices are incorporated into standard USEPA-type exposure and risk models to evaluate application and post-application exposures and risks for non-nutrient metals. By incorporating health-protective model parameters and assumptions, the outcome incorporates a margin of safety for the farm family, professional applicators and the general public. The risk assessments conclude that inorganic fertilizers are generally safe. Suggested 'safe limits' for 12 metals are provided in Table V. Manufacturers and distributors need to exercise responsible care and pay particular attention to micronutrient fertilizers containing recycled materials.

Introduction

The inorganic fertilizer industry has a long history of providing essential plant nutrients to meet the needs of agriculture. These include both primary nutrient (NPK) fertilizers as well as micronutrient fertilizers. In recent years, questions regarding the presence and safety of trace metal components in these fertilizer products have been raised in public forums. The same ore bodies that produce the essential plant nutrient elements (i.e., phosphorus, potassium, boron, calcium, chlorine, cobalt, copper, iron, magnesium, manganese, molybdenum, sodium and zinc) can also contain naturally occurring non-nutrient metals (e.g., arsenic, cadmium, lead). Another possible source of trace metals in fertilizers is recycled materials. These by-product materials are used because of their high content of essential nutrient metals (e.g., iron, manganese, and zinc). They account for a fraction of one percent of all inorganic fertilizers. While this recycling practice is allowable under existing laws, there is always the potential for less-than-desirable scenarios. In the interest of ensuring public health, some stakeholders have called for standards (numerical limits) for non-nutrient metals in inorganic fertilizers.

Chemical risk assessment is a widely used and accepted scientific practice that serves as a sound basis for setting health standards (i.e., safe limits) as part of the risk management process. Three stakeholders including the United States Environmental Protection Agency (U.S.EPA), the California Department of Food and Agriculture (CDFA) and The Fertilizer Institute (TFI) commissioned fertilizer risk assessments in response to the questions being raised. These three groups and their risk assessment experts worked cooperatively, sharing and reviewing each other's methodologies, input data and findings. All three risk assessments incorporated typical agricultural practices and high-end application rates covering a wide range of agricultural NPK and micronutrient fertilizers.

The three risk assessments, that is U.S. EPA, CDFA and TFI, share much of the same methodology and underlying science. All three conclude that, in nearly all cases, the metal levels found in fertilizers do not pose a risk to the applicators, to farm families or to the general public. The widely utilized 'risk paradigm' emphasizes use of current scientific thinking and modeling to regulate and/or otherwise manage chemicals in a health protective manner. The establishment of 'safe exposure levels' and 'safe levels of chemicals in products' is done using risk assessment techniques that take into account the inherent toxicity of a substance as well as the type and magnitude of exposure. These risk models are designed to assure the outcome is protective of health.

U.S.EPA's risk assessment entitled *Estimating Risk from Contaminants Contained in Agricultural Fertilizers* appeared as a draft report dated August 1999 (*1*). The stated primary purpose was 'to inform the Agency's decisions as to the need for federal regulatory action on fertilizer contaminants as well as to help guide state regulatory agencies making decisions about the need for and nature of risk-based standards for fertilizers'. U.S.EPA evaluated 9 metals in 13 fertilizer product categories used widely across the U.S. and concluded 'based on the data available, hazardous constituents in fertilizers generally do not pose harm to human health or the environment[1]'. The Agency did not recommend standards or limits for metals in fertilizers but tightened up a RCRA hazardous waste rule that allowed the use of zinc wastes as a recycled source of essential zinc nutrients in inorganic fertilizer. These wastes will be required to meet a technology-based metal limit in the future.

CDFA's risk assessment, entitled *Development of Risk-Based Concentrations for Arsenic, Cadmium and Lead in Inorganic Commercial Fertilizers*, was issued in March 1998 (*2*). Risk-based concentrations (RBCs) are calculated maximum levels (in parts per million – ppm) of a specific metal in a fertilizer product that do not pose a health risk following its use and exposure over a sensitive person's lifetime. The RBCs for the three metals have undergone some modification as a result of a scientific peer review of the 1998 assessment report. As a modification to the California Fertilizer Law and Regulations (Sections 2302 and 2303 of Title 3) effective January 1, 2002, CDFA's RBC values formed the basis for standards (limits) for arsenic, cadmium and lead in fertilizers. The numerical standards are not the RBC values per se, but take into account the results of the risk assessment as well as the political, economic and social aspects of the risk management decision making process specific to California.

TFI's risk assessment, entitled *Health Risk Evaluation of Select Metals in Inorganic Fertilizers Post Application*, was issued as a draft in January 2000 (*3*). RBCs were derived for 12 metals and compared to concentrations of these metals in NPK and micronutrient fertilizer products as reported by the U.S.EPA, numerous states monitoring programs and fertilizer manufacturers. Based on the same peer review comments as those received on the CDFA risk assessment (*2*), the RBC values have undergone some modifications and were finalized in July 2001 under the title *Scientific Basis for Risk-Based Concentrations of Metals in Fertilizers and Their Applicability as Standards* (see at www.tfi.org. or at www.aapfco.org).

Scope of the Fertilizer Risk Assessment

Chemical risk assessment is a widely used and accepted scientific approach

---

[1] U.S.EPA's screening-level assessment evaluated both human health and environmental (i.e., ecological) risks from fertilizer use. The CDFA and TFI assessments focused on human health risks.

to determine the magnitude of exposure (both frequency and duration) to a chemical and the likelihood that the exposure will exceed known hazard (toxicity) thresholds. Risk assessment can be used to predict whether a sensitive receptor (e.g., farm family adult and/or child) faces a health risk. The risk assessment for fertilizers is conducted at the screening level, meaning that the assumptions and default parameters used to describe exposure and toxicity are conservative (health protective) and will provide a margin of safety for the chosen health endpoint(s).

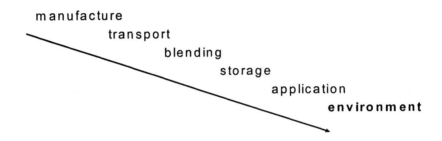

Figure 1 Lifecycle of a Fertilizer

As seen in Figure 1, there are a number of points along the life cycle of a fertilizer where exposures can occur. The risk assessment reported here focuses on the post-application, 'environmental' exposures of trace metals that remain in the soil or are taken up by the plant. TFI also commissioned a fertilizer risk assessment for applicators exposed to metals in fertilizers (4). TFI has recently completed a program to evaluate the exposure, hazards and potential risks associated with fertilizer materials and products (not just the metals) for the manufacturing and other pre-application stages of the fertilizer life cycle.

The fertilizer products with the highest potential for soil loading of metals are the focus of this assessment. These would include fertilizers with relatively high application rates and those with relatively high levels of trace metals. Much of the relevant published data, including fertilizer product compositions, application rates and metal content are in U.S.EPA's 1999 report entitled *Background Report on Fertilizer Use, Contaminants and Regulations* (5). Additional data reported by fertilizer manufacturers and by a number of state agencies were gathered and summarized. (6). Based on these available data, phosphate fertilizers were chosen to represent the macronutrient fertilizers (that is, nitrogen [N], phosphate [P], and potash [K]) and zinc, manganese, iron and boron fertilizers were chosen to represent the micronutrient fertilizers. Both groups were evaluated because while macronutrient products are generally applied at higher rates, some micronutrient products have higher concentrations of the trace metals.

Initial screening evaluations conducted by both U.S.EPA (*1*) and CDFA (*2*) identified the farm family as the receptor group having the highest potential for exposure and risk among all groups that might come in contact with metals from fertilizers. Therefore, risk-based protective levels for metals in fertilizer products that are based on protecting the farm family would also be protective for other receptors and exposure scenarios including, for example, professional applicators of fertilizer as well as home owners with a vegetable garden and/or who apply turf products. Efforts to derive risk-based concentrations (RBCs) for metals in fertilizer products therefore focused on the farm family.

Exposure to farm family members (children and adults) was estimated by combining route-specific exposures from dermal contact with fertilized soil, unintentional soil ingestion and ingestion of crops that may take up metals. Root crops, vegetables and grains were evaluated separately and in combination based on tendencies for metal uptake and on their presence in typical diets.

Twelve metals were evaluated in the TFI assessment including arsenic, cadmium, chromium, cobalt, copper, lead, mercury, molybdenum, nickel, selenium, vanadium and zinc. Other North American fertilizer evaluations, including U.S.EPA (*1,7*), CDFA (*2*) and the Canadian Fertilizers Act (R.S.1985, c.F-10), in total include these same 12 metals, but each has focused on different metals. The various factors and numerical values corresponding to children and adult receptors, to exposure scenarios, to soil and crop levels of metals and to the toxicity of the metals are described in the methodology section.

Narrowing the scope of a screening-level risk assessment to focus on fertilizer product types, exposure scenarios and metals posing the highest potential health risks is consistent with U.S.EPA risk assessment guidance (*8,9*). The assessment is thus designed to cover reasonable maximum likely exposures under the foreseeable range of typical use scenarios and agricultural conditions.

A conceptual model for this risk assessment is depicted in Figure 2. Looking at each component from left to right:

- Measured metal levels from a wide range of fertilizer products are available from the published literature, U.S.EPA, state agencies and fertilizer manufacturers for phosphate materials (e.g., diammonium phosphate [DAP]), for NPK products (e.g., a 10-5-5 blend) and for micronutrient products (e.g., providing boron, iron, manganese, or zinc).
- Fertilizer application rates are taken from published sources (rates used in the U.S.EPA (*1*), CDFA (*2*) and TFI (*3*) risk assessments are comparable).
- Soil accumulation of metals is estimated using U.S.EPA models and soil partition coefficient (K_d) values from the literature. TFI and CDFA assessments initially used published values from a 1983 review article by

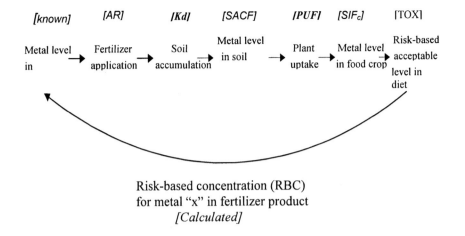

Figure 2. Conceptual Model for Fertilizer Risk Assessment

Baes and Sharp (10). CDFA's revised assessment uses values published by Sauve et al in 2000 (11). TFI's revised assessment uses the values proposed by U.S.EPA in the Agency's 1999 risk assessment (1).

- Metal levels in the soil are estimated using U.S.EPA soil accumulation models as described in CDFA's assessment (2). These levels can also be measured. There are some soil monitoring data and programs underway (e.g., a state-wide effort in California directed by UC Riverside scientists) to validate the model estimates. Soil levels are used for the dermal contact and unintentional ingestion exposure routes in the risk assessment.
- Plant uptake factors (PUF) are available in the literature; they vary by metal and by local conditions. TFI's assessment uses PUFs proposed by U.S.EPA (1) as well as those derived from the literature by Dr. Roland Hauck (a noted soil scientist formerly with the Tennessee Valley Authority) for metals where U.S.EPA did not provide an estimate. CDFA's contractor derived PUFs from the literature for their assessment.
- Metal levels in food crops are estimated using estimated soil levels multiplied by a conservative PUF. These levels are used for the dietary ingestion exposure route. The dietary route was determined to contribute the vast majority of the total exposure versus inhalation of particulates, dermal contact, and unintentional ingestion of soil.
- Risk-based protective levels in diet are established toxicity threshold levels (both cancer and non-cancer endpoints as appropriate). The assessment uses U.S.EPA-proposed no observed adverse effect levels (NOAEL).
- Risk-based concentrations (RBC) for a given metal in fertilizer are back-calculated from the protective level in the diet along with consideration of

concurrent direct exposures to the same metal in fertilizer and in fertilized soil. This 'unit' RBC is based on 1% of the nutrient (i.e., P_2O_5, boron, iron, manganese or zinc) and the corresponding application rate (AR) needed to achieve the desired amount of nutrient in the soil. RBCs are estimates of the maximum level of a metal in a fertilizer product that still assures its level in soil and in crops will not pose an unacceptable lifetime health risk.

- The RBC can then be compared to the measured levels of that metal in a product to determine if there is a likely health risk. The unit RBC must first be adjusted to account for the fraction of nutrient (FON) in the product of interest. [Recall the unit RBC is based on 1% FON.]

Risk Assessment Methodology

As previously noted, this is a screening-level risk assessment. It follows the generally accepted U.S.EPA standard approach of evaluating a reasonable maximum likely exposure under the foreseeable range of typical use scenarios and environmental conditions.

The human health risk equation is developed using standard U.S.EPA risk practices and exposure parameters modified to fit the fertilizer-in-soil scenario (8,9). The standard equation to calculate health risk combines three factors: the concentration of a specific metal in a fertilizer product, an estimated intake from exposure to that metal, and the chronic toxicity of that metal. Each metal is assessed separately. In keeping with standard U.S.EPA practices (and consistent with the U.S.EPA fertilizer risk assessment) a target cancer risk (TR) of 1×10^{-5} and a target non-cancer risk, or hazard quotient (THQ), of 1.0 is used. Following U.S.EPA guidance and the Agency's fertilizer risk assessment (1), the 90th percentile risk level is selected as the appropriate point estimate.

In the TFI and CDFA assessments, a back-calculated risk based approach is applied. That is, the equation is arranged to solve for a risk based concentration (RBC) of metal 'x' (say, As) in a NPK or micronutrient fertilizer product with 1% nutrient content. To apply the results to fertilizers in the marketplace, the RBC is adjusted to account for the actual nutrient level in a specific product (or product type). The adjusted RBC is then compared with the high-end measured concentration of the metal in that product. Coming at it from a slightly different vantage point, U.S.EPA (1) applied a forward-calculated risk based approach. The high-end of measured concentrations from among available samples of a specific type of fertilizer product is plugged into a risk equation to determine whether exposure to that metal poses a health risk. Both backward and forward approaches use the same general risk equation and protective risk levels. The forward calculation can be used for products in the marketplace today. The backward calculation can be used for products in the marketplace today as well as for future products. The backward calculation therefore provides a basis for establishing risk-based limits for current and future products.

RBC Equation

The RBC equation is presented below. The equation integrates the three potential routes of exposure; i.e., crop ingestion, unintentional soil ingestion and dermal contact with fertilized soil. Inhalation of soil particulates (dust) was not included because it represents such a low amount of potential exposure. Crop ingestion contributes by far the most to exposure.

$$RBC = \frac{TR\, or\, THQ}{SACF * \{AR * 1/FON * [(\frac{ED * EF * IRs * RAFs * CF}{BW * AT}) * TOX) + (\frac{ED * EF * SA * AF * ABS}{BW * AT} * TOX) + (\frac{ED * EF * IRc * RAFc}{BW * AT} * PUF * TOX)]\}}$$

where:

$\dfrac{ED * EF * SA * AF * ABS * CF}{BW * AT}$ = Summary Intake Factor (SIFd) Dermal Contact Soil

$\dfrac{ED * EF * IRs * RAFs * CF}{BW * AT}$ = Summary Intake Factor (SIFsi) Incedental Soil Ingestion

$\dfrac{ED * EF * IRc * RAFc}{BW * AT}$ = Summary Intake Factor (SIFc) Ingestion Crop

where:

RBC	=	Risk Based Concentration (mg MOPC/kg product);
TR/THQ	=	Protective Target Risk or Hazard Quotient (Unitless);
AR	=	Application Rate (g/m²-year);
FON	=	Fraction of Nutrient (e.g., P_2O_5) in product (unitless);
SACF	=	Soil Accumulation Factor (m²-year/g);
ED	=	Exposure Duration (years);
EF	=	Exposure Frequency (days/year);
BW	=	Body Weight (kg);
AT	=	Averaging Time (days);
CF	=	Conversion Factor (1x 10^{-6} kg/mg);
IRs	=	Ingestion Rate Soil (mg/day);
SA	=	Surface Area (cm²/event-day);
AF	=	Adherence Factor (mg/cm²);
IRc	=	Ingestion Rate Crops (kg/day);
RAF	=	Relative Absorption Factor (RAF) (unitless);
ABS	=	Dermal Absorption Factor (unitless);
PUF	=	Plant Uptake Factor (unitless); and
TOX	=	Toxicity Values (mg/kg-day or mg/kg-day^{-1}).

A separate RBC equation is applied for a multi-crop farm scenario where more than one crop type is grown within the year on the same field. The equation is generally the same as for the single crop farm, but all three-crop groups (i.e., roots, vegetables and grains) are integrated into one equation. The main difference is that each crop group has a different fertilizer application rate (AR) and plant uptake factor (PUF). There is also the addition of a new factor, fraction of land (FOL), in the equation that assigns the percentage or fraction of land for each of the three crop groups. Based on the CDFA assessment (2) the fractions assigned are 50% grain, 40% vegetable and 10% root for multi-crop scenarios. A 100% value is used in single-crop scenarios for a given crop type.

Numerical Values Used in the Equation

All numerical values used to derive RBCs are sumarized in Tables I and II. Table I includes numerical values for all parameters used to calculate summary intake factors (SIF). Table II includes numerical values for application rate (AR), fraction of land (FOL), soil accumulation (SACF), plant uptake (PUF), soil:water partition coefficient (K_d) and toxicity (TOX). Values for the SIF parameters are standard U.S.EPA health-protective default values (8, 12, 13). The toxicity values are also U.S.EPA recommended. Specific sources include IRIS, the Integrated Risk Information System (14) for all but two metals. HEAST, the Health Effects Assessment Summary Tables (15) is the source for vanadium and U.S.EPA (16) is the source for lead.

Unit RBC Calculation

Fertilizer products vary widely in the percent nutrient(s) they contain. For example, the granular fertilizer with the highest phosphate content is MAP which contains 52% P_2O_5; DAP is 46% P_2O_5 and so on. The P_2O_5 content in an NPK product varies with its intended use; it may be 10% (10-10-10) or 5% (25-5-5) or some other percentage. Similarly, the nutrient content in a micronutrient product (e.g., zinc or boron) can vary widely among products. The term used here to describe the nutrient content is 'fraction of nutrient' (FON). Because the nutrient content is varied among fertilizer products, the RBCs are developed on the basis of 1% of nutrient (FON = 1); that is, 1% P_2O_5 or 1% zinc, or boron, etc. The

Table I. Summary of the Parameters Used to Calculate Summary Intake Factors (SIFs)

Parameter	Units	Parameter Values	
Target Cancer Risk and Hard Quotient	unitless		
TR = Target Cancer Risk		1 in 100,000	
THQ = Target Hazard Quotient		1.0	
Biological Exposure Parameters		Adult	Child
EF = Exposure Frequency	days/year	350	350
ED = Exposure Duration	years	30	6
AT = Averaging Time	days		
Cancer		25,550	25,550
Noncancer		10,950	2190
BW = Body Weight	kg	71.8	15.5
IR_s = Soil Ingestion Rate	mg/day	50	200
IR_c = Ingestion Rate	g/kg-day		
Vegetables		1.7	2.9
Roots		1.1	2.1
Grains		3.4	9.4
AF = Adherence Factor	mg/cm^2	0.08	0.3
SA = Exposed Skin Surface Area	cm^2/day	5,700	2,900
RAF = Relative Absorption Factor	unitless	Soil	Crop
Arsenic		0.42	1
Cadmium		1	1
Chromium		1	1
Cobalt		1	1
Copper		1	1
Lead		0.41	0.5
Mercury		1	1
Molybdenum		1	1
Nickel		1	1
Selenium		1	1
Vanadium		1	1
Zinc		1	1
ABS = Dermal Absorption Factor	unitless		
Arsenic		0.03	
Cadmium		0.01	
Chromium		0.01	
Cobalt		0.01	
Copper		0.01	
Lead		1	
Mercury		0.01	
Molybdenum		0.01	
Nickel		0.01	
Selenium		0.01	
Vanadium		0.01	
Zinc		0.01	

Table II. Parameters Used to Calculate Risk Based Concentrations (RBCs)

Parameter	Parameter Values	
SACF = Soil Accumulation Factor	Using Baes & Sharp k_d and Hauck PUF	Using EPA k_d and EPA PUF
Arsenic	2.4	16.0
Cadmium	2.4	16.0
Chromium	16.0	17.0
Cobalt	11.0	11.0
Copper	6.9	15.0
Lead	30.0	66.0
Mercury	16.0	17.0
Molybdenum	6.5	8.7
Nickel	12.0	16.0
Selenium	1.0	4.6
Vanadium	3.9	9.7
Zinc	5.4	16.0

Units are m^2-yr/g; all values are 10^{-5}

AR = Application Rate	Vegetable	Root	Grain
Phosphate	13	17	6.9
Zinc Micronutrient	1.1	1.1	1.1

Units are g/m^2-yr; equivalent rates in lbs/acre are 119, 157, 63 for phosphate and 10 for micronutrients

PUF = Plant Uptake Factor (from Hauck 2000) – unitless	Vegetable	Root	Grain
Arsenic	0.03	0.0061	0.03
Cadmium	0.17	0.11	0.12
Chromium	0.00014	0.00018	0.037
Cobalt	0.003	0.0017	0.008
Copper	0.0034	0.027	0.31
Lead	0.008	0.0061	0.05
Mercury	0.061	0.082	0.26
Molybdenum	0.065	0.0111	0.12
Nickel	0.015	0.0086	0.05
Selenium	0.088	0.093	0.57
Vanadium	0.001	0.0005	0.001
Zinc	0.17	0.056	0.58

Plant Uptake Factor (continued) (from EPA 1999 unless noted)	Vegetable	Root	Grain
Arsenic	0.012	0.0026	0.014
Cadmium	0.119	0.07	0.084
Chromium	0.00011	0.0001	0.027
Cobalt (Hauck 2000)	0.003	0.0017	0.008
Copper	0.023	0.02	0.21
Lead	0.0049	0.0043	0.039
Mercury	0.016	0.0252	0.07
Molybdenum (Hauck 2000)	0.065	0.0111	0.12
Nickel	0.009	0.0054	0.04
Selenium	0.051	0.0471	0.38
Vanadium (Hauck 2000)	0.001	0.0005	0.001
Zinc	0.115	0.0383	0.45

Table II. Parameters Used to Calculate Risk Based Concentrations (RBCs) (Continued)

Parameter	Parameter Values		
FOL = Fraction of Land *Unitless*	Vegetable 0.4	Root 0.1	Grain 0.5
Kd = Soil-water Partition Coefficient	Lower-bound	Upper-bound	
Arsenic	6.7[a]	1,770[c]	
Cadmium	6.7[a]	957[c]	
Chromium	2,200[a]	6,981[c]	
Cobalt	55[a]	55[a]	
Copper	22[a]	310[c]	
Lead	99[a]	8,120[c]	
Mercury	330[b]	8,453[c]	
Molybdenum	20[a]	31[d]	
Nickel	63[b]	853[c]	
Selenium	2.7[a]	13[c]	
Vanadium	11[b]	39[e]	
Zinc	16[a]	530[c]	

Units are mL/g; [a] *Baes and Sharp 1983;* [b] *Gerritse et al. 1982;* [c] *EPA 1999;* [d] *Sauve et al. 2000 (pH 4.5-9);* [e] *Sludge Rule (Round 2)*

Toxicity Value	Oral	Dermal
SF = Slope Factor		
Arsenic	1.5	1.5
RfD = Reference Dose		
Arsenic	0.0003	0.00029
Cadmium	0.001	0.001
Chromium	1.5	0.03
Cobalt	0.06	0.026
Copper	0.04	0.039
Lead	see footnote	see footnote
Mercury	0.0003	0.000021
Molybdenum	0.005	0.005
Nickel	0.02	0.0001
Selenium	0.005	0.005
Vanadium	0.007	0.00021
Zinc	0.3	0.24

Units are (mg/kg-day)$^{-1}$ for slope factor and mg/kg-day for reference dose. Lead toxicity follows the USEPA approach and uses an acceptable blood lead level of 10 µg/dL.

resulting RBC values are designated 'unit RBCs'. A RBC for a specific product must therefore be adjusted to match the nutrient content (FON) of that product. For example, if the unit RBC for arsenic in phosphate fertilizers is "X" parts per million (ppm), then the RBC for a phosphate fertilizer with 10% P_2O_5 (or FON=10) would be 10X ppm. Fertilizers are added to the soil to deliver a desired amount of nutrient, for example, 100 pounds per acre (112 kg/ha) of phosphate (as P_2O_5). The desired amount of nutrient can be delivered as higher application rate of a low P_2O_5 content product or lower application rate of a high P_2O_5 content product. The calculation of a unit RBC for each metal allows its use in assessing health risks across a wide range of products and agricultural practices. This assessment established unit RBC values based on typical published fertilizer application rates (i.e., 119 lbs P_2O_5/acre [133 kg P_2O_5/ha], 157 [176] and 63 [71], for vegetable, root and grain crops, respectively, and 10 lbs micronutrient/acre [11.2 kg/ha] for all three crop types).

Most Sensitive Parameters in the Model

A simple sensitivity analysis was conducted to determine which of the variable parameters in the risk model has the greatest affect on the resulting RBC values. Many of the parameters are fixed in the risk equation, for example, exposure duration (ED) and frequency (EF), body weight (BW), crop (IR_c) and soil (IR_s) ingestion rates, absorption factors (ABS and RAF), and the toxicity values (TOX). The numerical values representing these parameters are well established in the literature and are those used by U.S.EPA (*1,8,9,12,13,14*). Whenever there was a range of possible values for a model parameter, a representative high-end value was chosen to provide a health protective risk prediction. Other parameters vary considerably depending upon local conditions. They include the soil adsorption coefficient (Kd) which has a major affect on the soil accumulation factor (SACF), the plant uptake factor (PUF), and the application rate of fertilizer (AR). These parameters are key components of the generalized risk model as depicted in Figure 2 [indicated in brackets and *italics*]. Kd and PUF have especially large influences on the resulting RBC values. The application rates are those used in US EPA's fertilizer risk assessment (*1*). As noted in Table II, reasonable AR values, taking into account geographic differences, were used in the risk calculation.

Outside reviewers indicated that the soil:water partitioning coefficients (K_d) used in CDFA's 1998 risk assessment (*2*) were probably too low for representing the range of K_d values across the range of environmental conditions. The values were from a 1983 review article (10) that summarized K_d values measured in the laboratory under conditions intended to mimic agricultural soil conditions. One reviewer pointed out that 'if the same K_d values used to predict soil accumulation and metal RBCs were used in soil-to-groundwater-leaching models, the prediction would be high metal levels in groundwater. And since high groundwater levels are not being observed, the K_d values used are unrealistically low.' U.S.EPA conducted a literature search of K_d values for their 1999 risk assessment. The data came from field as well as laboratory studies and they selected 'only K_d values derived for settings that most closely approximate the conditions found in agricultural soil...' [(*1*), Appendix D]. U.S.EPA's K_d database included a much wider range of values (for a given metal) than did those reported by Baes and Sharp (*10*). K_d values are highly dependent upon the form of the metal applied and environmental conditions such as soil type, pH, and organic matter. Table II provides the K_d values for each of the 12 metals used in this risk assessment along with their source. Note that, as was the case when using 'low end' K_d values in groundwater leaching models, if the 'high-end' K_d values reported in the U.S.EPA database are used in soil accumulation models, the prediction would be for very large buildup of these metals (10-100-1000X) in agricultural soil and corresponding increases in crop levels of these metals. Since the available data indicate this is not occurring in soil or in crops (e.g., *17, 18, 19, 20*), these 'high-end' K_d values are equally unrealistic. Reality apparently lies somewhere in between.[2]

The PUF is a measure of how much metal is taken up by a plant. PUFs reported in the literature are typically based on laboratory and/or greenhouse studies. At equilibrium, PUF = metal concentration in the plant ÷ metal concentration in the soil. Like K_d values, PUFs can vary widely depending upon the form of the metal applied and the local conditions such as soil type, pH, and organic matter content. U.S.EPA developed a database of PUF values as part of their 1999 fertilizer risk assessment (*1*). For metals not covered in the Agency's assessment, Dr. Roland Hauck (a noted soil scientist, now retired from the Tennessee Valley Authority) conducted a literature search and suggested PUF values based on available data and professional judgement. Table II provides the numerical PUF values for each of the 12 metals in this risk assessment along with their source. There is a widely recognized inverse correlation between a PUF and a K_d value for a given metal as depicted in Figure 3. When K_d is high,

---

[2] A 'low' K_d value equates to less metal found sorbed to the soil, more found soluble in the pore water surrounding the soil particles and therefore relatively more metal available to be taken up into plants as well as to leach toward groundwater. Conversely, a 'high' K_d value equates to more metal sorbed to soil, less in the pore water and therefore relatively less metal available to be taken up into plants but relatively more metal is likely to accumulate in the surface soil.

the corresponding PUF is low and when K_d is low, the corresponding PUF is high (*11*). This has to do with a metal's water solubility and availability to be taken up with the pore water by the roots, or conversely with a metal's affinity to sorb strongly to soil solids and not being available for plant uptake. In calculating RBC values for each of the metals, PUFs and K_ds are paired (higher-with-lower and vice versa but never low-with-low or high-with-high) in the soil accumulation and plant uptake estimations to account for the inverse correlation.

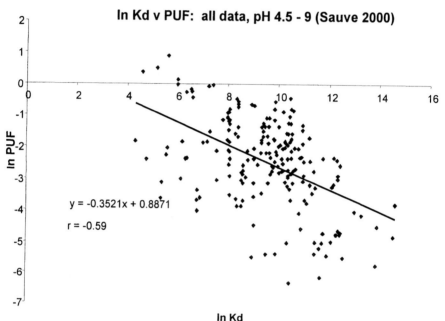

Figure 3. *Inverse relationship correlation between Kd's and PUF's*

RBCs for 12 Metals in Fertilizers

The relatively large effect that the choice of K_d values has on the resulting RBC values is obvious. There are valid arguments to support that the lowest K_d values reported in the literature are unrealistically low when used to predict the uptake of metals into plants. Likewise there are valid arguments to support that the highest K_d values are unrealistically high. The most scientifically defensible RBCs are somewhere in between (i.e., bounded by) these two extremes. Two sets of RBCs were therefore calculated, one with the Baes and Sharp (*10*) K_d values and the second with the U.S.EPA (*1*) K_d values. All other parameters and model components were kept the same. The K_d : PUF inverse correlation was also applied in coming up with 'the most representative' RBCs.

RBCs are calculated for both phosphate and micronutrient fertilizers, for both adult and child farm residents, and for each of three crop types separately and for a multi-crop scenario (50% grain, 40% vegetable and 10% root). Note that dermal contact with fertilized soil as well as unintentional ingestion of fertilized soil are included in the total exposure, however the predominant exposure comes from the metal in the crop that is consumed.

The unit RBCs, based on a 1% FON, are presented in Table III (using Baes and Sharp K_ds) and Table IV (using U.S.EPA K_ds). The lowest unit RBC for each metal (the far right column in Tables III and IV) would be a health protective value under all reasonable foreseeable scenarios. The lowest RBCs are for the child for all metals except arsenic. Arsenic is a carcinogen and therefore the adult farm resident is at greater risk because the exposure duration is much longer for an adult. The RBC values in Table IV are lower than the corresponding values in Table III. This is because the K_d values used to generate the RBC values in Table IV are proportionally higher than those used to generate the RBC values in Table III. Given the design of the screening-level risk model and the numerical values incorporated into the model, both sets of RBCs are considered health protective. Each is more or less conservative than the other but both are health protective.

The risk assessors who prepared the CDFA risk assessment and those who prepared the TFI assessment agree that the midpoint value between the upper-bound and lower-bound RBCs for a given metal is a scientifically defensible screening-level value. This concept is depicted in Figure 4. The example is for arsenic in phosphate fertilizer. The lower bound conditions (i.e., U.S.EPA K_d values and U.S.EPA and/or Hauck PUFs) result in a unit RBC of 7 mg/kg arsenic. The upper bound conditions (i.e., Baes and Sharp K_d values and Hauck PUFs) at the 90 percentile risk level result in a unit RBC of 19 mg/kg arsenic. The midpoint value is 13 mg/kg (7 + 19 ÷ 2 = 13). Midpoints can be calculated for all 12 metals using the values in the far right columns in Tables III and IV. The resulting RBC values in ppm (ppm = mg/kg) are shown in Table V.

Again, these RBC values are 'unit RBC values' meaning they are based on 1% FON in the product. In order to apply these unit RBC values to any given product, the fraction of nutrient (either P_2O_5 for a phosphate product or zinc, boron, etc. for a micronutrient product) is multiplied times the unit RBC value. For example, a phosphate product containing 10% P_2O_5 would equate to a 10 X 'unit RBC for arsenic' limit for that product (i.e., 10 X 13 ppm = 130 ppm).

Table III. Risk Based Concentrations (RBCs) for all Scenarios (Upper Bound Conditions)

Metal	Adult Farm Resident RBC				Child Farm Resident RBC				Lowest Unit RBC
	Vegetable	Roots	Grains	Multi-crop	Vegetable	Roots	Grains	Multi-crop	
Phosphate Fertilizer									
Arsenic	99	210	24	19	190	310	38	34	19
Cadmium	120	250	52	36	68	130	21	16	16
Chromium (III)	1,800,000	1,400,000	110,000	100,000	170,000	130,000	36,000	34,000	34,000
Cobalt	45,000	66,000	12,000	8,400	15,000	15,000	4,000	3,100	3,100
Copper	70,000	11,000	820	760	20,000	5,000	300	280	280
Lead	1,900	1,200	190	170	1,300	840	110	97	97
Mercury	15	13	3.3	2.2	7.8	6.1	1.2	0.9	0.9
Molybdenum	330	2,200	160	100	180	900	56	42	42
Nickel	3,800	5,400	1,400	1,000	990	930	450	350	350
Selenium	2,600	2,900	380	300	1,500	1,400	140	120	120
Vanadium	36,000	29,000	10,000	8,300	4,200	3,200	2,800	2,200	2,200
Zinc	15,000	53,000	4,200	3,100	8,800	25,000	1,500	1,200	1,200
Micronutrient Fertilizer									
Arsenic	1,300	2,800	180	150	2,400	4,100	280	260	150
Cadmium	1,400	3,200	380	320	840	1,800	150	130	130
Chromium (III)	21,000,000	21,000,000	670,000	670,000	2,100,000	2,100,000	220,000	220,000	220,000
Cobalt	540,000	1,000,000	73,000	62,000	180,000	230,000	25,000	23,000	23,000
Copper	830,000	180,000	5,200	5,000	230,000	78,000	1,900	1,800	1,800
Lead	24,000	17,000	1,400	1,400	16,000	12,000	770	740	740
Mercury	170	200	21	17	93	95	7.4	6.5	6.5
Molybdenum	3,900	35,000	980	760	2,200	14,000	350	300	300
Nickel	46,000	84,000	9,000	7,500	12,000	14,000	2,900	2,600	2,600
Selenium	31,000	46,000	2,400	2,100	17,000	22,000	870	800	800
Vanadium	420,000	450,000	64,000	59,000	50,000	51,000	17,000	17,000	17,000
Zinc	180,000	830,000	26,000	23,000	100,000	400,000	9,600	8,600	8,600

Units for all values are mg/kg (or ppm)

Table IV. Risk Based Concentrations (RBCs) For All Scenarios (Lower Bound Conditions)

Metal	Adult Farm Resident RBC				Child Farm Resident RBC				Lowest Unit RBC
	Vegetable	Roots	Grains	Multi-crop	Vegetable	Roots	Grains	Multi-crop	
Phosphate Fertilizer									
Arsenic	39	13	17	7	55	22	27	14	7
Cadmium	33	69	13	9.1	20	37	5.3	4.1	4.1
Chromium (III)	1,800,000	1,400,000	140,000	140,000	170,000	130,000	46,000	43,000	43,000
Cobalt	82,000	120,000	29,000	20,000	20,000	18,000	9,100	6,800	6,800
Copper	5,100	7,000	550	460	2,500	2,900	200	180	180
Lead	470	990	49	45	350	600	27	25	25
Mercury	49	37	11	7.5	21	15	4.0	3	3
Molybdenum	410	2,700	210	130	230	950	74	54	54
Nickel	4,000	4,800	1,400	1,000	830	720	420	330	330
Selenium	1,000	1,300	130	100	540	590	45	40	40
Vanadium	15,000	13,000	24,000	12,000	1,700	2,300	3,000	1,900	1,900
Zinc	7,600	27,000	1,900	1,400	4,300	12,000	670	560	560
Micronutrient									
Arsenic	490	180	120	71	710	300	190	130	71
Cadmium	410	930	94	79	250	510	38	33	33
Chromium (III)	21,000,000	22,000,000	900,000	900,000	2,100,000	2,100,000	290,000	290,000	290,000
Cobalt	980,000	1,800,000	180,000	150,000	230,000	280,000	57,000	51,000	51,000
Copper	61,000	110,000	3,400	3,200	30,000	45,000	1,200	1,200	1,200
Lead	5,900	13,000	370	350	4,400	8,100	190	180	180
Mercury	580	580	71	58	250	230	25	22	22
Molybdenum	4,900	42,000	1,300	1,000	2,700	15,000	460	390	390
Nickel	48,000	75,000	8,600	7,500	9,900	11,000	2,600	2,400	2,400
Selenium	12,000	20,000	790	720	6,500	9,200	290	270	270
Vanadium	180,000	200,000	150,000	110,000	21,000	21,000	19,000	17,000	17,000
Zinc	91,000	420,000	12,000	10,000	51,000	190,000	4,200	3,800	3,800

Units for all values are mg/kg (or ppm)

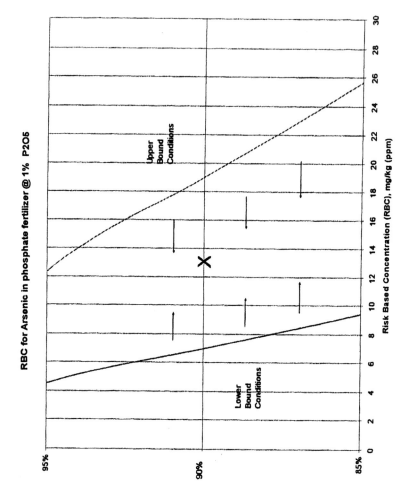

Figure 4. Example RBC Depiction for As in Fertilizer

Table V. Unit RBC Values for Metals in Fertilizers

	ppm per 1% P_2O_5	ppm per 1% micronutrient
Arsenic	13	112
Cadmium	10	83
Cobalt	3,100	23,000
Lead	61	463
Mercury	1	6
Molybdenum	42	300
Nickel	250	1,900
Selenium	26	180
Zinc	420	2,900

Use of RBCs in Evaluating Product Safety and in Setting Standards for Metals in Fertilizers

Risk assessment is the prevailing scientific standard method to judge health and environmental safety of chemicals. Risk assessment provides a basis for risk management decisions including whether a particular product is safe and/or in setting standards (or limits) for chemicals as components of products.

U.S.EPA's 1999 fertilizer risk assessment (*1*) was designed to evaluate the safety of fertilizer products in use today, based on their metal content. The Agency's assessment uses the standard, screening-level exposure and health risk model to determine whether published measured levels for nine metals pose health risks. U.S. EPA evaluated 13 product categories (roughly 260 samples) and concluded 'based on the data available, hazardous constituents in fertilizers generally do not pose harm to human health or the environment'.

The risk assessment described herein was designed to derive RBCs for 12 metals in phosphate and micronutrient fertilizer products. Any fertilizer can be evaluated for safety with regards to its metal content using the unit RBC values as a screening tool. For example, if arsenic is kept at or below 13 mg/kg in phosphate fertilizer (i.e., 13 ppm) for each 1% of P_2O_5, the fertilizer can be used safely, that is, its use will not pose a lifetime health risk for the most highly exposed individuals. Therefore, a product with a 10-10-10 composition would need to contain less than 130 ppm arsenic (13 x 10) to pass the RBC screen.

In a situation where a single product contains both P_2O_5 and micronutrients, multiply the percent P_2O_5 by the appropriate RBC value and multiply the percent micronutrient (use the highest percentage if more than one micronutrient is present in the product) by the appropriate RBC value. The higher of the two multiplication products, that is, based on either P_2O_5 or on the micronutrient, becomes the limit for that metal in that fertilizer product. This approach works because the product would be applied either at therate typical for the P_2O5 nutrient or the micronutrient, and bothe scenarios are covered in the respective

RBC calculations. Consider, for example, an agricultural blend product that is 10-10-10 plus 5% zinc. The arsenic limit would be 560 ppm (10 X 13 = 130 and 5 X 112 = 560).

The available data on measured metal levels in fertilizers was assembled as part of this risk assessment project. The data sources include the same database that U.S.EPA used in their 1999 assessment (1) as well as a survey of fertilizer manufacturers conducted in 1999 (6) and data provided by state monitoring programs (1996-2000). A total of approximately 1050 samples of phosphate fertilizers, NPK blends and micronutrient products were identified. The range and average level for each metal are presented in Table VI. The maximum level for each of the 12 metals was compared to its corresponding RBC (adjusted for the FON in the sample tested). In all but a handful of samples, the measured level of metal was below the screening RBC value thus indicating that those fertilizers did not pose a health risk for the most highly exposed individuals. All of the phosphate containing fertilizers pass the screen, as do 10 of the 12 metals in all fertilizer samples. The only exceptions are samples of a few micronutrient products where arsenic and lead exceed the screening RBC values. This very closely parallels the results reported in U.S.EPA's 1999 risk assessment (1) and the Agency's overall conclusion as quoted above. As stated at the outset, a screening level risk assessment incorporates assumptions that will result in a conservative (protective) outcome that provides a margin of safety for the chosen health endpoint(s).

The RBCs derived in this assessment have also been used in a standard setting procedure. In 2002, the midpoint RBC values for nine of the metals were adopted as interim standards by the Association of American Plant Food Control Officials (AAPFCO) as guidance for states that may decide to set limits on levels of trace metals in fertilizers (see at www.aapfco.org).

Summary

The post-application health risk assessments conducted by U.S.EPA, CDFA and TFI are the first comprehensive assessments for fertilizers. The conceptual models, the methodologies and key parameter values, and the results are very consistent reflecting the use of a fairly standardized, screening-level risk-based approach. These assessments support a sound science-based conclusion that fertilizers in the marketplace are safe under typical use conditions. This 'safety' extends to those who apply the fertilizers (at home, commercially, and farm families) as well as to the general public who consume food grown using fertilizers. Such is clearly the case for products that are sold to deliver the primary nutrients nitrogen, phosphate and potash (NPK) as well as for the vast majority of the micronutrient products. Available product monitoring data

Table VI. Metal Concentrations in Phosphate and Micronutrient Fertilizer Products

Metal	Minimum mg/kg	Maximum mg/kg	Mean mg/kg
Phosphate Fertilizer			
Arsenic	0.05	155	10
Cadmium	0.015	250	13
Chromium	0.25	5,060	120
Cobalt	0.04	58	5.6
Copper	0.14	1,170	14
Lead	0.05	5,425	13
Mercury	0.001	1.5	0.16
Molybdenum	0.69	72	12
Nickel	0.5	351	22
Selenium	0.03	27	2.6
Vanadium	0.28	1,106	128
Zinc	0.30	6,270	260
Micronutrient Fertilizer			
Arsenic	0.1	6,200	400
Cadmium	0.095	3,900	120
Chromium	0.25	8,100	290
Cobalt	0.25	790	200
Copper	0.5	40,000	7,700
Lead	0.32	52,000	2,400
Mercury	0.0025	12	1
Molybdenum	0.25	850	83
Nickel	0.5	8,950	88
Selenium	0.013	25	6
Vanadium	0.5	47	23
Zinc	6	350,000	120,000

indicate that the highest measured levels of arsenic and lead in a few micronutrient products, specifically those that incorporate recycled materials as sources of essential plant micronutrients, exceed the screening-level RBC values. While this does not mean that these few products are universally unsafe, it points to a need for closer case-specific evaluations and for marketplace vigilance.

While exposures at relatively low levels of nearly all substances can occur with negligible health risk, exposure at some relatively high level to those same substances can result in adverse effects. Simply put, 'the dose does make the poison'. Additional work is underway to further refine the scientific basis of these fertilizer risk assessments. The various stakeholders are measuring levels of metals in products and monitoring levels of metals in soils and crops in areas where fertilizers are applied at high rates. The RBC values in Table V (once adjusted for product specific FON) can be used with a high degree of confidence to screen fertilizer products. Manufacturers and distributors of inorganic fertilizers (particularly micronutrient fertilizers that contain recycled materials) that contain 'high' levels of trace metals should exercise responsible care.

References

1. U. S. Environmental Protection Agency (USEPA); Estimating risks from contaminants contained in agricultural fertilizers; draft; Office of Solid Waste and Center for Environmental Analysis; Washington, DC, 1999.
2. Development of risk-based concentrations for arsenic, cadmium, and lead in inorganic California Department of Food and Agriculture and the Heavy Metal Task Force (CDFA); commercial fertilizers; Foster Wheeler Environmental Corporation; Sacramento, CA, 1998.
3. The Weinberg Group Inc. (TWG); Health risk evaluation of select metals in inorganic fertilizers post application; draft, January 2000.
4. The Weinberg Group Inc. (TWG); Health Risk Based Concentrations for Fertilizer Products and Fertilizer Applicators; February 1999.
5. U. S. Environmental Protection Agency (USEPA); Background report on fertilizer use, contaminants and regulations; Battelle Memorial Institute; Columbus, OH, 1999b.
6. The Weinberg Group Inc. (TWG); Industry and literature survey for nutritive & non-nutritive elements in inorganic fertilizer materials; prepared for The Fertilizer Institute; Washington, DC, 1999c.

7. U. S. Environmental Protection Agency (USEPA); A guide to the biosolids risk assessments for the EPA Part 503 rule; EPA832-B-93-005; Office of Wastewater Management; Washington, DC, 1995.
8. U. S. Environmental Protection Agency (USEPA); Risk assessment guidance for superfund; Volume I; Human health evaluation manual (Part A); interim final; EPA/540/1-89/002; Office of Emergency and Remedial Response; Washington, DC, 1989.
9. U. S. Environmental Protection Agency (USEPA); Guidelines for exposure assessment; Office of Health and Environmental Assessment; Washington, DC, 1992.
10. Baes, C.F.; Sharp, R.D. *J. Environ. Qual.* 1983, 12, 17-28.
11. Sauve, S.; Hendershot; W.; Allen, H.E. *J. Environ. Sci. Technol.* 2000, 34:7, 1125-1131.
12. U. S. Environmental Protection Agency (USEPA); Exposure factors handbook; EPA/600/P-95/002FA,b,c; Volumes I, II, and III; Office of Research and Development; Washington, DC, 1997.
13. U. S. Environmental Protection Agency (USEPA); Risk assessment guidance for superfund; Volume I.; human health evaluation manual; supplemental guidance; dermal risk assessment; draft; NCEA-W-0364; Office of Emergency and Remedial Response; Washington, DC, 1998.
14. U. S. Environmental Protection Agency (USEPA); *Integrated risk information system-IRIS*; Office of Research and Development; National Center for Environmental Assessment, 1999, www.epa.gov/iris.
15. U. S. Environmental Protection Agency (USEPA); Health effects assessment summary tables [HEAST]; EPA540-R-97-036; Office of Solid Waste and Emergency Response; Washington, DC, 1997b.
16. U.S. Environmental Protection Agency (USEPA); Recommendations of the technical review workgroup for lead for an interim approach to assessing risks with adult exposures to lead in soil, 1996.
17. Johnston, A.E.; Jones, K.C. *The Fertilizer Society Proceedings*; The origin and fate of cadmium in soil, 1995.
18. Gunderson, E.L. *J. of AOAC Inter.* 1995, 78, No. 4, 910-921.
19. Hodson, M.E.; Valsami-Jones, E.; Cotter-Howells, J.D. *J. Environ. Sci. Technol.* 2000, 16, 3501-3507.
20. Hettiarachchi, G.M.; Pierzynski, G.M.; Ransom, M.E. *J. Environ. Sci. Technol.* 2000, 34, 4614-4619.

Measurement, Impact, and Management of Fertilizer Nutrients

Chapter 11

Inorganic Nutrient Use in the United States: Past and Present

W. M. Stewart

Great Plains Director, Potash and Phosphate Institute, 3206-B 66th Street, Lubbock, TX 79493

Inorganic nutrient use in the US is affected by many factors. The impact of some of these factors is more easily quantified than others. For example, the effect of factors such as crop and fertilizer prices and area in production can be evaluated using historical records and current statistics. Other factors such as changes in crop genetics and management practices are more difficult to quantify. An evaluation of estimated nutrient removal/use ratio has revealed that the US is currently depleting P and K from soils on a national basis at an increasing rate each year. Over the 40-year period evaluated, soil K has been in the draw-down mode every year, and P has been in the depletion mode since the early 1980s.

Introduction

Inorganic nutrient use in the US has increased markedly over the past 40 years. **Figure 1** shows total consumption of nitrogen (N), phosphate (P_2O_5), and potash (K_2O) fertilizer from 1961 to 2000. The consumption curve in **Figure 1** consists of three distinct segments. The first segment represents a period of linear increase in consumption from 1961 to about 1974. The next segment is a period of erratic consumption extending from about 1974 to 1986. Fertilizer use has been relatively flat in the final segment from about 1986 to 2000.

Inorganic nutrient use in the US is influenced by many factors. Some of these factors are easily evaluated while others are not. Examples of factors that can be readily evaluated include area in production, fertilizer and crop price, government programs, and weather. Other factors affecting fertilizer consumption that may be more difficult to quantify include adoption of new technology, crop genetic improvements, and market development efforts.

Selected factors affecting inorganic nutrient use

The effect of crop acres (planted) on fertilizer consumption from 1961 to 2000 can be seen in **Figure 2**. Surprisingly, there was relatively little change in total area planted during the period of linear growth in fertilizer consumption (1961 to 1974). The relatively flat area planted from about 1961 to about 1972 suggests somewhat stable government programs in agriculture during this period. However, at the beginning of the period of erratic fertilizer consumption a more direct relationship between area in production and fertilizer use develops. The effect of area in production on fertilizer use is much more apparent in this period than previously. The largest shifts in area planted were in 1983 with the PIK (Payment in Kind) program and again from 1986 to 1987 with the CRP (Conservation Reserve Program). These government programs in turn had a substantial impact on $N+P_2O_5+K_2O$ fertilizer use. Since the CRP program there has been relatively little change in area planted and in inorganic nutrient consumption.

The impacts of the index of prices paid by farmers for fertilizer and the index of prices paid to farmers for crops on fertilizer consumption are shown in **Figures 3 and 4**. During the period of linear increased in fertilizer use (1961 to 1974) both crop and fertilizer prices were relatively stable. The stability in these factors and in crop acres during this period suggest a time of greater sense of security and confidence on the part of farmers to predict changes in factors affecting their operations (e.g., crop price, fertilizer price, and government programs). However, in the early to mid 1970s dramatic swings in both price indexes and consequently in fertilizer consumption were experienced. Factors

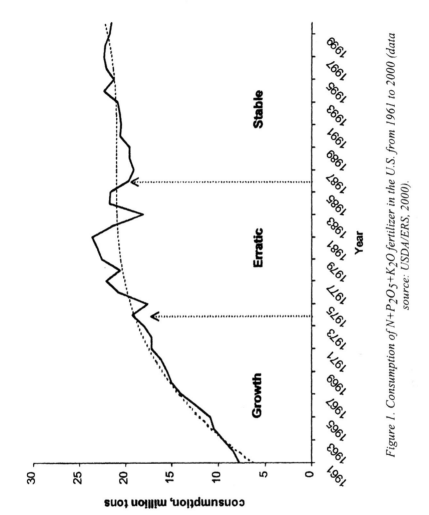

Figure 1. Consumption of N+P$_2$O$_5$+K$_2$O fertilizer in the U.S. from 1961 to 2000 (data source: USDA/ERS, 2000).

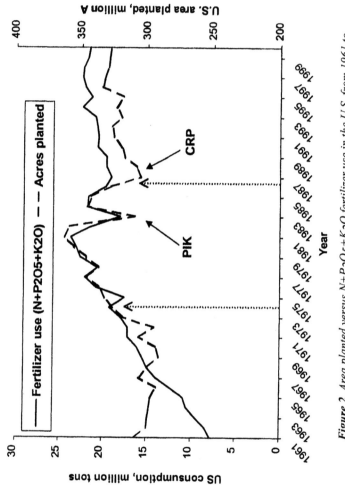

Figure 2. Area planted versus $N+P_2O_5+K_2O$ fertilizer use in the U.S. from 1961 to 2000 (data source for area planted: USDA/NASS, 2001. Historic Track Records, May 2001).

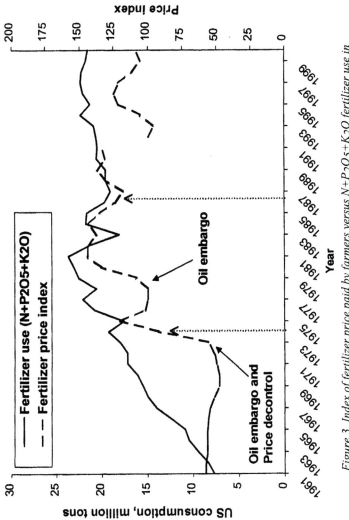

Figure 3. Index of fertilizer price paid by farmers versus $N+P_2O_5+K_2O$ fertilizer use in the U.S. from 1961 to 2000 (data source for price index: for 1961 to 1991, USDA/ERS, 1994, for 1992 to 2000, USDA/NASS, 2001).

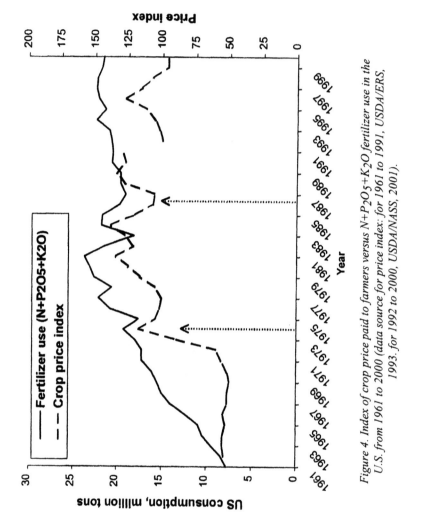

Figure 4. Index of crop price paid to farmers versus $N+P_2O_5+K_2O$ fertilizer use in the U.S. from 1961 to 2000 (data source for price index: for 1961 to 1991, USDA/ERS, 1993. for 1992 to 2000, USDA/NASS, 2001).

such as the oil embargo and fertilizer price decontrol of 1973 and another oil embargo in 1979 dramatically affected fertilizer price. In the early 1990s the method of calculating both indexes was changed, therefore the peaks from the early 1990s forward are not on the same scale as those in previous years.

Nutrient removal/use ratio

One way to gauge nutrient use is to consider it relative to nutrient removal. An attempt to estimate yearly national P and K removal/use ratio was made using production data from 18 crops in the US. From these production data total nutrient removal was estimated for each of 40 years from 1961 to 2000. The 18 crops represented an average of 98 percent of acres harvested. Therefore, practically all P and K removal nationwide was taken into account. Assumptions were made concerning the amount of P and K removed per unit of production for each crop. The yearly inorganic nutrient use data was the same as was used in **Figure 1**. **Figure 5** shows the estimated national removal/use ratio for both P and K from 1961 to 2000. During the time of linear increase in fertilizer consumption (1961 to 1974) the ratios of removal/use declined for both P and K, however during the period of erratic consumption (1974 to 1986) the ratios began to turn upward and have been steadily increasing since. This agrees with trends in yield of major crops and total P and K use. Over the entire 40-year period yields of major crops have increased. Both P and K consumption were increasing until around the mid 1970s. Phosphorus use in the US began to flatten and even trend downward in the mid to late 1970s while K use began to flatten in the early 1980s. Therefore, since the mid 1970s to early 1980s the US has been removing P and K from its soils at a steadily increasing rate. In fact, considering the 1:1 line in **Figure 5** the US has been in a depletion mode for soil K over the entire 40 years and has been depleting soil P since the early 1980s.

The removal/use ratios in **Figure 5** do not take into account organic nutrient use. However, only a small percentage of cropland actually receives nutrients from manure. For the four major US crops, the average percent of acres receiving manure from 1990 to 1997 was 17 for corn, 6 for soybeans, 4 for cotton, and 3 for wheat (USDA/ERS, 2000). Therefore, the increasing rate of depletion of soil P and K across the nation applies to the majority of acres in production.

The trends observed in **Figure 5** are not sustainable. As some point, if these trends continue, a flattening and possibly a reduction in yields of major crops will be observed. While soils in many major production regions have tremendous capacity to buffer the effect of drawdown of soil P and K on crop yield and performance, at some point even the most fertile soils will cease to produce increasing yields without adequate and balanced P and K input. A case

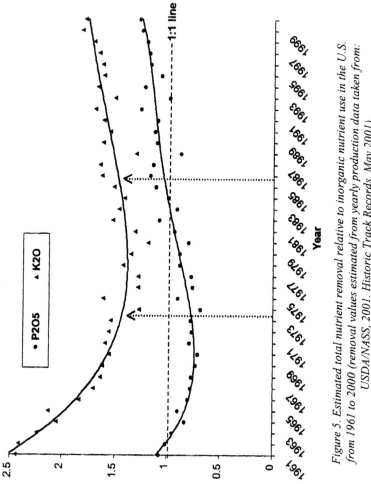

Figure 5. Estimated total nutrient removal relative to inorganic nutrient use in the U.S. from 1961 to 2000 (removal values estimated from yearly production data taken from: USDA/NASS, 2001. Historic Track Records, May 2001).

in point involves the increased incidence of K deficiency in corn and soybean production in the Corn Belt and the North Central regions of the US in recent years (Murrell, 2001). Another example is the increased occurrence of K deficiency in cotton throughout much of the cotton belt (Oosterhuis, 1994). These observations are likely a reflection of the increased removal relative to use of P and K. A decline in removal/use ratios by increasing the use of P and K relative to crop removal will avert the certain negative consequences of prolonged "mining" of these nutrients from US soils.

Conclusion

Inorganic nutrient use in the US is affected by many factors. The impact of some of these factors is more easily quantified than others. Fertilizer consumption in the US has dramatically increased over the past 40 years, however, this increase has not been consistent across years. From 1961 to about 1974 there was a period of linear increase in N+P+K consumption. From the mid 1970s until about 1986 there was a period of dramatic swings in fertilizer use. Finally, from 1986 to present consumption has been relatively flat. The effect of factors such as crop and fertilizer prices and area in production can be seen using historical records and current statistics.

An evaluation of estimated nutrient removal/use ratio has revealed that the US is currently in the "mining" mode for both soil P and K. According to this analysis the US is depleting P and K from soils on a national basis at an increasing rate each year. Over the 40-year period evaluated, soil K has been in the draw-down mode every year, and P has been in the depletion mode since the early 1980s. These trends cannot continue over the long term without consequences. These consequences will most likely come in the form of increased occurrence of P and K deficiency symptoms and eventual flattening and/or reduction in average yields of major crops. To avoid these consequences the removal/use ratio for P and K must be reduced. One method of achieving this, and thus avoiding eventual yield reductions, is to increase P and K use relative to removal.

References

Murrell, T.S. 2001. North Central Region Observations: Recognizing potassium deficiencies [Online]. Available at
http://www.ppi-ppic.org/ppiweb/usanc.nsf/$webindex/E536946145BEE7BF8625696500645065 (verified 26 Sept. 2001)

Oosterhuis, K.M. 1994. Potassium nutrition of cotton in the USA, with particular reference to foliar fertilization. p. 133-146. *In* G.A. Constable and N.W. Forrester (eds.) Challenging the Future: Proceedings of the World Cotton Research Conference-1. Brisbane, Australia. 1994.

USDA/ERS. 1993. Agricultural Outlook Yearbook [Online]. Available at. http://usda.mannlib.cornell.edu/usda/ (verified 23 Sept. 2001).

USDA/ERS. 1994 Fertilizer Use and Price Statistics [Online]. Available at http://usda.mannlib.cornell.edu/ (verified 23 Sept. 2001).

USDA/ERS. 2000. Agricultural Resources and Environmental Indicators, 2000 [Online]. Available at http://www.ers.usda.gov/Emphases/Harmony/issues/arei2000/ (verified 23 Sept. 2001).

USDA/NASS. 2001. Agricultural Prices 2000 Summary [Online]. Available at. http://usda.mannlib.cornell.edu/usda/ (verified 23 Sept. 2001).

USDA/NASS. 2001. Historic Track Records [Online]. Available at http://www.usda.gov/nass/pubs/trackrec/track01d.htm (verified 23 Sept. 2001).

Chapter 12

Documenting Nitrogen Leaching and Runoff Losses from Urban Landscapes

J. L. Cisar[1], J. E. Erickson[1], G. H. Snyder[2], J. J. Haydu[3], and J. C. Volin[4]

[1]University of Florida, Fort Lauderdale Research and Education Center, 3205 College Avenue, Fort Lauderdale, FL 33314 (email: jcli@ufl.edu; telephone: 954–577–6336; fax: 954–475–4125)
[2]Department of Soil and Water Science, University of Florida, Everglades Research and Education Center, P.O. Drawer A, Belle Glade, FL 33430
[3]Mid-Florida REC, 2705 Binion Road, Apopka, FL 32703
[4]Division of Science, Florida Atlantic University, 2912 College Avenue, Davie, FL 33314

Urbanization and land use changes near coastal areas have been shown to degrade water quality. In an effort to reduce nitrogen (N) pollution from urban areas, various programs are promoting alternative landscape materials, which require less fertilizer N inputs than traditional turfgrass vegetation. Although plant materials that require less N input may conceivably reduce N pollution from urban landscapes, this paradigm remains unresolved. Here we use a replicated field experiment to show that nitrogen pollution via leaching was significantly greater on a mixed-species ornamental landscape compared to a turfgrass monoculture, despite half the nutrient input over a one-year period. Our results indicate that turfgrass is relatively efficient at using applied nitrogen and when properly maintained offers minimal environmental impact.

Introduction

Urban areas with large population concentrations have the potential to strongly impact water resources throughout the world (*1*). Residential and commercial landscapes associated with urban areas have some potential for loss of nitrogen (N) in surface runoff and leaching due to the predominance of intensely maintained turfgrass areas. For example, the rise in nitrate-N levels at Weeki Wachee Springs in Hernando County, Florida, appears to be highly correlated with population growth in the county (*2*). In fact, over 50% of waters in Florida are affected by urban nonpoint-source pollution, which includes residential landscapes (*3*). As a consequence, considerable interest is now being focused on quantifying the loss of fertilizer N applied to residential landscapes through runoff and leaching.

Nitrogen is the nutrient applied to turfgrass in the greatest quantity and frequency (*4*). For example, St. Augustinegrass (*Stenotaphrum secundatum* (Walt.) Kuntze), the most common turfgrass for residential lawns in Florida, is a moderate fertility warm season grass that receives 150 to 300 kg N ha^{-1} yr^{-1} when appropriately fertilized (*4*). The fate of fertilizer N applied to residential landscapes involves gaseous loss to the atmosphere through volatilization and denitrification, plant uptake, soil storage, runoff to surface water, and leaching to ground water. The major inorganic forms of N found in the soil are ammonium-N (NH_4-N) and nitrate-N (NO_3-N). While both inorganic forms can be lost through runoff and leaching, an emphasis is generally placed on NO_3-N because of its mobility and biological activity. Since human consumption of NO_3-N can be deleterious, the United States Environmental Protection Agency has established a drinking water standard of 10 mg L^{-1} (*5*). In addition, coastal estuaries and bays, which have been found to be N limited, may be degraded by NO_3-N concentrations lower than the drinking standard (*6*). Therefore, residential landscape management practices that minimize fertilizer N runoff and leaching are advantageous to both human safety and the environment.

Several factors have been shown to affect surface runoff and leaching including: 1) vegetation type and density; 2) fertilizer source and rate; 3) frequency and intensity of precipitation event; 4) soil properties; and 5) slope (*7*). A number of authors have examined N runoff and leaching from turfgrass systems, which is the predominant component of conventional residential landscapes (*4,8-10*). In a two year runoff study conducted in the northeastern United States, Morton et al. (*9*) observed only two surface runoff events from a cool season turfgrass lawn in which < 7% of applied fertilizer N was lost from any treatment. In both cases, the runoff events were attributed to unusual climatic conditions. Similarly, Gross et al. (*11*) reported very little nitrogen runoff from turf when compared with agronomic row crops. Still, the potential for surface runoff from residential landscapes exists, especially from vegetation on compact soils that receive intense rainfall (*12-13*). However, surface runoff and sediment losses from turfgrass are generally thought to be relatively low, likely a result of the high infiltration capacity and vegetation density associated with most turfgrass species (*14-15*).

Conversely for N leaching, several studies have described conditions where fertilizer N losses to ground water from turfgrass systems have occurred through leaching (*9,16-17*). Generally, management practices that included readily-available N fertilizers, such as NH_4NO_3 or urea, and frequent or excessive irrigation were found to facilitate the leaching of applied nutrients, especially on sandy soils. For example, Snyder et al. (*17*) reported that 56% of applied N was lost through leaching from bermudagrass grown on sandy soils using NH_4NO_3 for fertilizer with daily irrigation and rainfall. Petrovic et al. (*18*) observed up to 47% of N leached when applied as urea N from Kentucky bluegrass grown on a sandy loam soil with no irrigation. However, in the same study no N leaching was observed from ureaformaldehyde or Milorganite N sources applied at the same rates. Therefore, it appears that judicious management practices can greatly reduce N leaching. Still N leaching from turfgrass is a major concern. In residential areas where turfgrass is a major land use, NO_3-N leaching from turfgrass has been proposed as a significant source of nitrate contamination of ground water (*19*).

As a result of the relatively high maintenance requirements of many traditional landscape materials, a number of authors have proposed the use of alternative plant materials in residential landscapes, which require minimal fertilizer and supplemental irrigation to be maintained in a healthy state (*13, 20*). While landscapes using these alternative plant materials are generally perceived to require less water and fertilizer inputs, few studies anywhere have examined the loss of fertilizer N applied to alternative landscapes. Hipp et al. (*13*) examined the use of resource efficient plants for reducing chemical and nutrient runoff, but came to no generalization with respect to landscape effect on N loss because of conflicts with management practices and other variables. Reinert et al. (*21*) observed less runoff from xeriscapes grown on a silty clay soil, which was attributed to the level of irrigation practiced (antecedent soil moisture). Based on similar resource efficient principles, the Florida Yards and Neighborhoods (FYN) program was established in 1992. Partially in response to concerns over nonpoint-source pollution from residential landscapes, the FYN program advocates the use of alternative landscape materials that require fewer inputs and may provide additional environmental benefits over conventional turfgrass lawns (*22*). Landscapes utilizing the principles of the FYN program are intended to enhance the environment by reducing harmful runoff and providing wildlife habitat. Further considerations include aesthetics, food production, climate control, and resale value (23). Although the FYN landscapes offer a wide variety of potential environmental benefits, a major emphasis is placed on reducing N loading to ground and surface waters (*22*). However, the FYN program has no N runoff and leaching data from FYN landscapes to support their principles. Furthermore, no N runoff or leaching data is available for St. Augustinegrass, the predominant turfgrass used for residential landscapes in Florida.

Because nonpoint-source pollution is a pervasive and severe problem in southern Florida as well as throughout the world, a field-scale facility was constructed to monitor runoff and percolate from two properly managed contrasting landscape types. The objective of this study was to compare fertilizer N losses via surface runoff and leaching from a high maintenance St. Augustinegrass monoculture versus a lower maintenance mixed-species landscape employing principles espoused by the FYN program.

MATERIALS AND METHODS

Construction of Experimental Facility. A facility containing eight 9.5x5m research plots was constructed at the University of Florida's Fort Lauderdale Research and Education Center to collect both surface runoff and subsurface percolate from two contrasting landscape treatments (24). One treatment consisted of a St. Augustinegrass monoculture (*Stenotaphrum secundatum* (Walt.) Kuntze cv. 'Floratam') and the other treatment was an arrangement of ornamental ground covers, shrubs and trees that followed the principles of the FYN program. The St. Augustinegrass was maintained at a height of 7.5 cm. The clippings were removed for the first six months of the study and mulched for the final six months. The mixed-species landscape consisted of twelve ornamental species and contained no turfgrass. Eucalyptus mulch was uniformly applied at a depth of 7.5 cm on the mixed-species landscape to reduce soil water evaporation and weed growth. Mr. Allen Gardner of the FYN program chose the plant materials and developed the design for the mixed-species landscape (Fig. 1).

Construction of the facility commenced in the fall of 1998. A crushed limestone foundation layer provided a 10% slope (Fig. 2). The root zone mix used was a medium-fine sand (33.4 and 54.9% respectively) characteristic of residential sandy soils in Florida (Boynton Sand and Gravel, Palm Beach County, Florida), which have relatively high infiltration rates (total pore space = 37.8%). At the time of soil installation the soil pH was alkaline (pH > 8.0), however when tested at the end of the study the pH across all plots was near neutral ranging from 6.6 – 7.4.

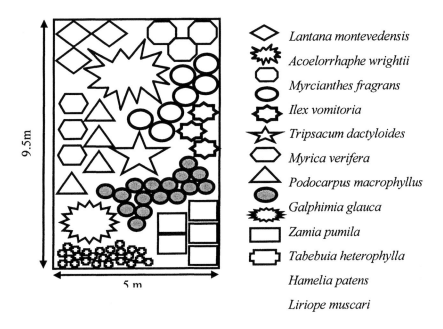

Figure 1. Layout of the mixed-species landscape. The diagram shows the quantity and relative position of all species included within each 9.5x5m plot.. The apex of the 10% slope coincides with the top of the diagram.

Maintenance. In each of the eight plots we installed a rectangular shaped perimeter irrigation system comprised of six inward-facing spray nozzles, providing uniform irrigation. After a 5-month establishment period, the irrigation on the ornamental plots was converted to a microjet irrigation system, which directed water to the plants. An irrigation time clock controlled each plot as a separate zone. An automatic rain shutoff switch was connected to the time clock to avoid irrigation following sufficient rainfall. The pressure on the irrigation was maintained at approximately 210 kPa. Daily irrigation was recorded based on a flow meter installed in the irrigation system. Soil percolate was measured initially by random manual measurements on each plot and subsequently by tipping bucket flow gauges (Unidata America-Model 6406H). Rainfall was recorded continually and averaged monthly.

Both landscape treatments were maintained according to general recommendations for residential landscapes in Florida to obtain information on the effect of landscape type on N leaching. The rate of N was 5.0 g N m^{-2} per application for both treatments. However, the fertilizer was applied every two

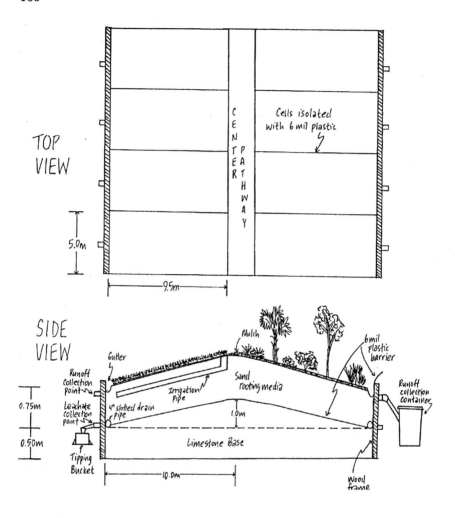

Figure 2. The 20x20m facility constructed to assess surface runoff and leaching from two contrasting landscape types. Eight plots (9.5x5m) were created, which allowed for four replications of each treatment. A plastic barrier (6-mm) provided hydrological isolation for each plot. A gutter system collected any surface runoff, while 10.2-cm slotted pipes drained the percolate. Each landscape was established in 0.75 m of medium-fine sand at a 10% slope.

months to the St. Augustinegrass (300 kg N ha^{-1} yr^{-1}) and every four months to the ornamentals (150 kg N ha^{-1} yr^{-1}). Three fertilizer cycles, corresponding with fertilization dates on the mixed-species landscape, were established to facilitate data analysis (Table 1). A blended 26-3-11 (N-P$_2$O$_5$-K$_2$O) granular fertilizer (LESCO Inc., Sebring, FL) was chosen for both treatments, except for the last cycle on the mixed-species landscape whereby a 12-2-14 (N-P$_2$O$_5$-K$_2$O) mix was used to supply more potassium and micronutrients to the ornamental species. According to the fertilizer bag label in Florida, nitrogen sources in the fertilizer were urea (58%), sulfur-coated urea (37.5%), and ammonium phosphate (4.5%). The granular material was hand distributed and watered in with approximately 5.0 mm of irrigation at each application. Planting of both treatments occurred on December 18, 1998. The first fertilizer cycle and data collection commenced in February 1999 (Table 1). Nitrogen leaching and runoff data were collected for a twelve-month period following the onset of fertilization.

Table I. Fertilizer cycles. The dates given are the days fertilizer was applied to the respective treatment. Each application was applied at a rate of 5 g N m^{-2}.

Cycle	Mixed-species	St. Augustinegrass
1 (Feb – May)	2/4/99	2/4/99
		4/5/99
2 (Jun – Sep)	6/2/99	6/2/99
		8/3/99
3 (Oct – Jan)	10/7/99	10/7/99
		11/29/99

Water Sample Collection and Chemical Determination. Initially, soil water (leachate) samples along with percolate flow measurements were taken at least once daily from a slotted drainage pipe placed across the lower edge of each plot, which drained the percolate for the entire plot (Fig 2). Beginning in July, a percolate sample was collected every six hours and pooled into a daily composite sample using ISCO (model 2900) auto-samplers. Both runoff and percolate water samples were immediately acidified with sulfuric acid upon collection and refrigerated at 4 °C until analysis. The samples were analyzed for inorganic N (NH$_4$-N and NO$_3$-N) using colorimetric auto-analyzers (OI Analytical, EPA Methods 350.1 and 353.2). All analyses were performed in accordance with Florida Department of Environmental Protection guidelines at the University of Florida Analytical Research Laboratory (ARL) in Gainesville.

The nutrient loadings (quantity leached) were calculated by multiplying the concentration of each nutrient found in the daily composite sample by the volume of percolate measured for the respective 24h period. In addition, turf-clipping sub-samples were collected on each mowing occasion. The tissue samples were wet digested (26) and analyzed for N at the ARL in Gainesville, Florida.

Statistical Analysis. The experimental design for this study was a completely randomized design with a single factor, landscape type. The design included two treatments and four replications. Mean treatment effects were determined for nutrient runoff and leaching and evapotranspiration according to the three four-month fertilizer cycles. Statistically significant treatment effects on N runoff, N leaching, and ET were identified using procedures for ANOVA.

RESULTS AND DISCUSSION

Nitrogen Contributions from Surface Runoff. The rapid infiltration rate of the sandy soil (59.9 cm hr^{-1}) was most likely the major factor influencing the amount of surface runoff observed. Despite the 10% slope used on each plot, runoff was collected only once during the study after an unusually intense rainfall event (22 cm) in June. A hurricane may have produced negligble runoff in October but the strong winds hindered reliable data collection. In addition, the nutrient concentrations in the June runoff event were insignificant. No significant ($P < 0.05$) treatment difference was observed in the quantity of runoff and less than 0.175 cm was collected from each plot. Consequently, it appears that for both landscape types surface runoff occurs only under extreme conditions. The insignificant treatment differences of inorganic-N concentrations observed on the June runoff event corroborate with other previous work that found very little N loss via runoff from residential landscape systems (14, 16). In a recent study on a fine-textured soil where runoff was observed, turfgrass was found to reduce concentrations of sediment and associated contaminants in runoff, but not volume of runoff (27). Schmitt et al. (27) also reported that young trees and shrubs planted in grass filter strips provided no benefit over the grass alone. These results suggest that turfgrass may be appropriate for reducing the loss of sediment and associated contaminants in surface runoff, although it may have no effect on soluble nutrients such as inorganic N from fertilizers. Thus, while no differences were observed in our study, it is possible that differences in sediment and sediment bound N may occur in runoff events between contrasting landscape types.

Percolate Nitrogen Contributions from Rainfall. The landscapes received 205.4 cm of rainfall during the twelve-month period of data collection (Fig 3).

Southern Florida generally experiences a wet season (June – November) and a dry season (December – May).

During our study 183.8 cm of rain was received in the wet season and only 21.6 cm in the dry season. Annual measurements of percolate were 223.7 and

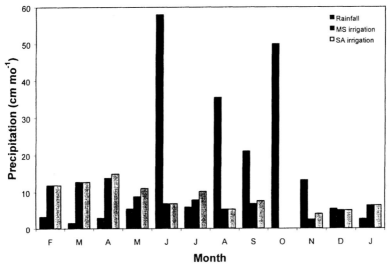

Figure 3. Monthly precipitation inputs. Over the duration of the study 8.3% less irrigation was applied to the mixed-species (MS) treatment compared to the St. Augustinegrass (SA) treatment. Irrigation was measured by a flow meter in the main irrigation line.

208.2 cm on the mixed-species and turfgrass landscapes, respectively (Fig. 4). Rainfall had the largest effect on the volume of percolate observed as considerable amounts of percolate were measured during the rainy months of the wet season. For instance, the impact of rainfall events can be clearly seen when comparing the months of June (wet season) and December (dry season) of 1999. Approximately 59.0 cm of percolate was measured in June and only 5.6 cm in December. Accordingly 58.0 and 5.5 cm of rainfall were received in June and December, respectively. Therefore, the large volume of rainfall driven percolate in the wet season is likely responsible for transporting mobile ions such as NO_3-N into ground water. Rainfall may also offset the decreased nutrient leaching achieved through efficient irrigation. In addition, rainfall may contribute nutrients to the landscapes through wet deposition. The 1.23 mg L^{-1} of inorganic N (0.55 and 0.68 mg L^{-1} for NO_3-N and NH_4-N respectively) found in the rainfall was similar to levels previously reported in southern Florida. A recent USGS water quality assessment report found mean concentrations of total nitrogen ranging from 0.77 to 1.45 mg L^{-1} in wet and dry atmospheric precipitation at seven sites in southern Florida from 1990 - 1992 (27). Thus, based on 205.4 cm of rainfall received during the study and a mean concentration of 1.23 mg L^{-1}, an additional 2.5 g N m^{-2} was potentially received

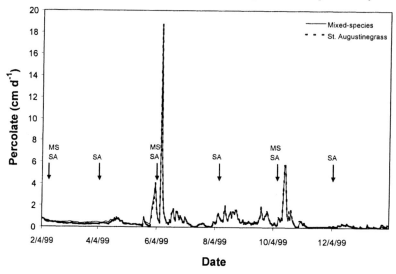

Figure 4. Volume of percolate measured daily. Lines represent mean treatment values, n = 4. Arrows indicate fertilization events for the St. Augustinegrass (SA) and the mixed-species landscape (MS).

via rainfall. A nominal 1.09 mg L^{-1} of inorganic N (0.07 and 1.02 mg L^{-1} for NO_3-N and NH_4-N respectively) found in irrigation water as well as nitrogen from rainfall is made available in small quantities throughout the year and is likely incorporated biologically, having little impact on groundwater.

Water Budget Summary A water budget summary including rainfall, irrigation, percolate and ET was created for each of the fertilizer cycles (Table 2). Although Florida usually receives a substantial amount of rainfall, irrigation is frequently still required to maintain healthy, attractive residential lawns because of the high temperatures and soils with a relatively low water holding capacity. Since the turfgrass generally experienced water stress more rapidly than the mixed-species landscape, it received 8.3% more irrigation over the year (95.1 and 87.2 cm of irrigation were applied to the grass and ornamentals, respectively). Over half of the irrigation (approximately 53%) was applied during the first cycle (February - June) for both treatments because the plant materials were not established and less than 20 cm of rainfall was received during the period. Significantly greater percolate was measured from the mixed-species treatment during the first cycle, while ET was significantly greater for the St. Augustinegrass during the same period. However, no significant differences in percolate or ET were seen between the treatments in the two

Table II Water budget (cm) summary for the three 4-month cycles. ET was calculated using the following formula: ET = Rainfall + Irrigation − Percolate (17). Values are treatment means based on n = 4. Rainfall, irrigation, and percolate were all measured parameters at the site.

Cycle (4 mo)	Treatment	Rainfall	Irrigation	Percolate	ET
1 (Feb-May)		19.44			
	Mixed-species		47.36	61.17a*	5.63a
	Turfgrass		50.66	48.30b	21.80b
2 (Jun-Sep)		115.85			
	Mixed-species		27.07	116.08a	24.84a
	Turfgrass		30.12	118.08a	29.88a
3 (Oct-Jan)		70.09			
	Mixed-species		14.05	46.47a	42.35a
	Turfgrass		15.58	41.79a	39.20a

*Values within a column and cycle followed by the same letter are not significantly different at $P < 0.05$.

subsequent cycles. A similar trend was observed upon examining monthly ET, where a relative monthly difference is observed the first 5 or 6 months, after which no differences are seen. As expected, ET increases during the warm, wet months and decreases during the cool, drier months. During the dry season (December - May), ET was approximately 4.5 to 5 cm per month on the grass and on the ornamentals following establishment. ET was more variable during the wet season (June - November) averaging approximately 10 cm per month for both landscape treatments. The large peak in October is likely a slight overestimate of ET due to an underestimation of percolate during a heavy rain event in which monitoring equipment was flooded. The initial differences in ET were likely the result of the longer establishment requirements of the ornamental species compared to the relatively rapidly establishing turfgrass. Presumably, reduced root development and reduced canopy on the ornamental plants were less efficient at transpiring water during establishment. Similar ET rates between treatments during the second two cycles were likely due to increased transpiration as the ornamentals became more established. By measuring the water holding capacity of new eucalyptus mulch (6.6% on a volume basis), it was determined that approximately 0.5 cm of uniformly applied irrigation could be absorbed by a dry 7.5-cm mulch layer and never reach the soil. Thus, 20 – 40% of routine applied irrigation was potentially absorbed by the mulch and evaporated. Therefore, it was unlikely that the transpiration rates between the two landscapes were the same. Evaporation was probably greater on the mixed-species treatment because of the mulch layer. Irrigation applied below the mulch layer could alleviate this unnecessary loss of water.

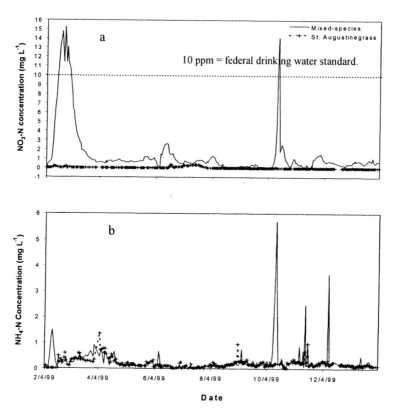

Figure 5a and b. (a) Nitrate nitrogen (NO_3-N) and (b) Ammonium nitrogen (NH_4-N) in the percolate water collected from the contrasting landscapes. The figure presents mean treatment values determined from a daily (24h) composite sample, n = 4. Data from 10/13 - 10/17 was based on n = 1 for both treatments.

Effects of Fertilization and Precipitation events on Nitrogen. The daily average concentrations of NO_3-N and NH_4-N are presented in Figures 5a and b for the year. The average NO_3-N concentration found in the percolate from the mixed-species plots exceeded the 10 mg L^{-1} federal drinking water standard on several occasions. The mean NO_3-N concentrations in the percolate from the mixed-species landscape ranged from < 0.2 to 15.2 mg L^{-1} with a mean annual concentration of 1.46 mg L^{-1}. In contrast, the mean concentration of NO$_3$-N found in the percolate from the turfgrass was generally below 0.2 mg L^{-1} and never exceeded 0.4 mg L^{-1}, resulting in a mean annual concentration < 0.2 mg L^{-1}. Moreover, little variability in NO$_3$-N concentrations was observed following fertilization and rainfall events. Unlike NO$_3$-N, the concentrations of readily exchangeable NH$_4$-N appear to follow no real pattern with relationship to fertilization and precipitation events. The mean NH$_4$-N concentrations in the mixed-species percolate ranged from < 0.3 to 5.7 mg L^{-1} with an annual mean concentration of 0.3 mg L^{-1}, while on the St. Augustine grass the annual mean concentration was < 0.3 mg L^{-1} with a range of < 0.3 to 1.4 mg L^{-1}.

Consistent with the N concentration data of the percolate, the quantity of NO$_3$-N leached was also significantly greater from the mixed-species landscape compared to the St. Augustinegrass (Figs. 6a and b). The loss of NO$_3$-N from the mixed-species landscape was related to the application of fertilizer with three large peaks following fertilization events. In addition, there appears to be a relationship between rainfall and NO$_3$-N loss through leaching as relatively more peaks can be seen during the wet months, coinciding with heavy precipitation events. Therefore, rainfall and time since fertilization affected both the percolate concentrations of NO$_3$-N and especially the quantity of N leached from the mixed-species treatment. However, there appears to be no relationship between N leaching and rainfall or application of fertilizer on the St. Augustinegrass (*8,29*). As before, the quantity of NH$_4$-N leached appears to follow no real pattern, other than modest peaks following fertilization on the mixed-species landscape. The relatively low contribution of NH$_4$-N to the overall quantity of N leached from the mixed-species treatment was likely the result of rapid conversion to NO$_3$-N (study conditions favorable for nitrification) and the fact that NH$_4$-N is retained on the exchange complex of the soil (*7*).

Nitrogen Leaching Losses per Fertilization Cycle. N loading to groundwater was partitioned into three, four-month fertilizer cycles, which correspond with each application of fertilizer to the mixed-species landscape. Significant differences ($P < 0.001$) in N loss were observed between treatments during each of the three cycles (1.82, 1.26 and 1.72 g m^2 from the mixed-species and 0.12, 0.20 and 0.09 g m^2 from the turfgrass, respectively). For the 12 month duration of the study, mean leaching losses of inorganic N (NO$_3$-N + NH$_4$-N) amounted to 0.414 g N m^{-2} on the St. Augustinegrass compared to 4.891 g N m^{-2} on the mixed-species treatment. Thus, 1.4% and 32.6% of applied fertilizer N was leached from the grass and the mixed- species treatments, respectively. Approximately, 87.1% of the inorganic N leached from the St. Augustinegrass was ammoniacal in nature compared to just 17.3% on the mixed-species. Nitrogen losses were similar within landscape during each of the three fertilizer cycles. However, very little rainfall and percolate were measured during the first cycle when compared to the third cycle. Thus, there was a greater potential for leaching in the third cycle, yet the quantity of N leached was essentially the

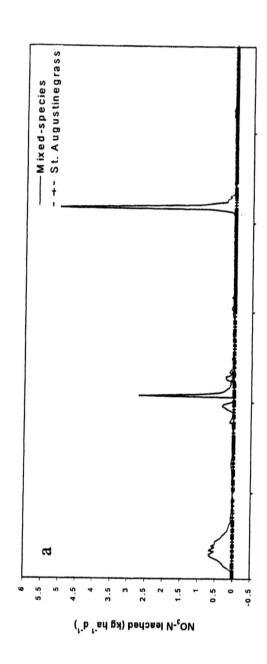

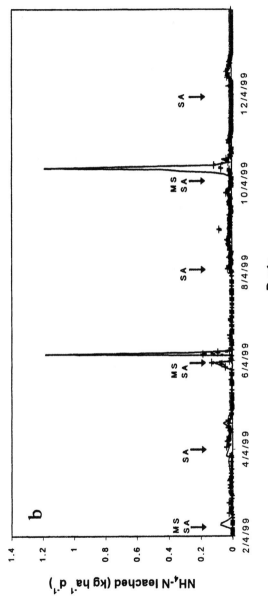

Figure 6a and b. Quantity of (a) nitrate nitrogen (NO3-N) and (b) ammonium nitrogen (NH4-N) leached daily. Lines represent mean treatment values, n = 4. Arrows indicate fertilization events for the St. Augustinegrass (SA) and the mixed-species landscape (MS). Data from 10/13 -10/17 based on n = 1 for both treatments.

same as the first cycle. N leaching may be greatly reduced as the mixed-species landscape develops.

A number of environmental conditions as well as management practices seem likely to have contributed to the results found in this study. As mentioned earlier, monthly rainfall was positively correlated with the amount of percolate measured ($R^2 = 0.86$, $P > 0.05$). The increase in percolate volume generally led to increased N leaching on the mixed-species treatment. Furthermore, the timing of precipitation in relation to fertilization also seemed to affect N leaching. Within ten days of fertilization during both of the last two cycles, storm events brought substantial rainfall that likely exacerbated the loss of N on the mixed-species landscape. These climatic conditions (> 20 cm of rainfall per day) occur sporadically in subtropical climates such as in southern Florida. However, even under these intense rainfall events very little N leaching was observed on the St. Augustine grass despite receiving twice as much fertilizer nitrogen.

Since very little N leaching or runoff was observed on the St. Augustinegrass and greater than 25% was lost from the mixed-species, it appears that applied N behaved quite differently between the two landscape treatments. By taking subsamples of the turfgrass clippings for tissue N analysis, it was determined that approximately 25% of applied N was incorporated into foliage. Similarly, Starr and DeRoo (8) reported that about 29% of applied fertilizer N was found in the turfgrass plant (clippings, shoots, and roots). Although roots and shoots were not sampled, both incorporated some N, yet there was still a large percentage of N unaccounted for in the St. Augustinegrass treatments. Since conditions were favorable for ammonia (NH_3) volatilization, such as an initially high pH and high temperatures, some N may have been lost to the atmosphere by volatilization (*30*). However, management practices that were used in this research such as slowly available N sources and irrigation after fertilization have been shown to greatly reduce volatilization (31). Starr and DeRoo (*8*) found between 36 to 47% of applied fertilizer N was stored in the soil-thatch pool. Assuming little volatilization on the mixed-species treatment, less N (as a percent of applied fertilizer N) was accounted for in vegetative uptake or soil storage and more lost via leaching compared to the turfgrass treatment.

CONCLUSIONS

Despite the 10% slope used for the plots, no significant surface runoff was observed from either landscape treatment throughout the duration of the study. Nutrient losses via surface water runoff appear unlikely from either treatment, perhaps due to the physical properties of the soil used in the study.

Significantly greater nitrogen leaching occurred from the mixed-species landscape compared to the St. Augustinegrass given the management principles used in this study. This seems to indicate that the St. Augustinegrass treatment was more efficient at utilizing applied nitrogen. While it is difficult to predict nitrogen leaching from either treatment over time, it appears that St.

Augustinegrass may be a more appropriate landscape type for reducing nitrogen contamination of ground water.

ACKNOWLEDGMENTS

We thank Karen Williams, Kevin Wise and Scott Park for technical support in the field and David Rich for laboratory assistance. The authors would also like to acknowledge the anonymous reviewers for their comments. This work was supported by the Florida Department of Environmental Protection, Sarasota Bay National Estuary Program, and the University of Florida Agricultural Experiment Station.

REFERENCES

1. Niedzialkowski, D. and D. Athayde. 1985. Water quality data and urban nonpoint source pollution: the nationwide urban runoff program. Pp. 437-441. *In* Perspectives on Nonpoint Source Pollution, Proc. National Conf., May 19-20,1985, Kansas City, MO.
2. Champion, K. 1999. Unpublished data. Southwest Florida Water Management District. Tampa, Florida.
3. Association of State and Interstate Water Pollution Control Administrators. 1984. America's clean water: The state's evaluation of progress 1972-982: appendix. Washington, D.C.
4. Cisar, J. L., G. H. Snyder, and P. Nkedi-Kizza. 1991. Maintaining quality turfgrass with minimal nitrogen leaching. Univ. of Fla., Inst. of Food and Agr. Sci. Bul. 273. 11 pp.
5. U. S. Environmental Protection Agency. 1976. Quality criteria for water. U.S. Gov. Print Office, Washington, D.C.
6. Ryther, J. A., and W. M. Dunstan. 1971. Nitrogen, phosphorus, and eutrophication in the coastal environment. Science (Washington, D.C.) 171:1008-1013.
7. Brady, N. C. 1990. The nature and properties of soils. Macmillan Publishing Company, New York. Tenth ed. 621 pp.
8. Starr, J. L., and H. C. DeRoo. 1981. The fate of nitrogen applied to turfgrass. Crop Sci. 21:531-536.
9. Morton, T. G., A. J. Gold, and W. M. Sullivan. 1988. Influence of overwatering and fertilization on nitrogen losses from home lawns. J. Environ. Qual. 17(1):124-130.
10. Petrovic, A. M. 1990. The fate of nitrogenous fertilizers applied to turfgrass. J. Environ. Qual. 19:1-14.
11. Gross, C. M., J. S. Angle, and M. S. Welterlen. 1990. Nutrient and sediment losses from turfgrass. J. Environ. Qual. 19:663-668.
12. Moe, P. G., J. V. Mannering, and C. B. Johnson. 1967. Loss of fertilizer nitrogen in surface runoff water. Soil Sci. 104:389-394.

13. Hipp, B., S. Alexander, and T. Knowles. 1993. Use of resource-efficient plants to reduce nitrogen, phosphorus, and pesticide runoff in residential and commercial landscapes. Water Sci. Tech. 28(No. 3-5):205-213.
14. Kelling, K. A. and A. E. Peterson. 1975. Urban lawn infiltration rates and fertilizer runoff losses under simulated rainfall. Soil Sci. Soc. Amer. Proc. 39:348-352.
15. Krenitsky, E. C., M. J. Carroll, R. L. Hill, and J. M. Krouse. 1998. Runoff and sediment losses from natural and man-made erosion control materials. Crop Sci. 38:1042-1046.
16. Brown, K. W., J. C. Thomas, and R. L. Duble. 1982. Nitrogen source effect on nitrate and ammonium leaching and runoff losses from greens. Agron. J. 74:947-950.
17. Snyder, G. H., B. J. Augustin, and J. M. Davidson. 1984. Moisture sensor-controlled irrigation for reducing N leaching in Bermudagrass turf. Agron. J. 76:964-969.
18. Petrovic, A. M., N. W. Hummel, and M. J. Carroll. 1986. Nitrogen source effects on nitrate leaching from late fall nitrogen applied to turfgrass. P. 137. *In* Agronomy abstracts. ASA, Madison, WI.
19. Flipse, W. J., Jr., B. G. Katz, J. B. Lindner, and R. Markel. 1984. Sources of nitrate in ground water in a sewered housing development, central Long Island, New York. Ground Water 32:418-426.
20. Sacamano, C. M. and W. D. Jones. 1975. Native trees and shrubs for landscape use in the desert southwest. Univ. AZ Coop. Ext. Serv. Bull. A-82. 40 pp.
21. Reinert, J. A., S. J. Maranz, B. Hipp, and M. C. Engelke. 1997. Prevention of pollution in surface runoff from urban landscapes. Pp. 7-31. *In* The Pollution Solution: Be Water Wise Proceedings. Texas A&M Univ. Dallas, Texas.
22. Best, C. H. 1994. Florida Yards and Neighborhoods program. Proc. Fla. State Hort. Soc. 107:368-370.
23. Garner, A., J. Stevely, H. Smith, M. Hoppe, T. Floyd, and P. Hinchcliff. 1996. A guide to environmentally friendly landscaping - Florida Yards and Neighborhoods handbook. U. of Florida Institute of Food and Ag. Sciences Bulletin 295. 56 pp.
24. Erickson, J. E., J. C. Volin, J. L. Cisar, and G. H. Snyder. 1999. A facility for documenting the effect of urban landscape type on fertilizer nitrogen runoff. Proc. Fla. State Hort. Soc. 112:266-9.
25. Hummel, N. W., Jr. 1993. Laboratory methods for evaluation of putting green root zone mixes. USGA Green Section Record. Mar/Apr. Pp. 23-33.
26. Lowther. 1986. Use of a single H_2SO_4-H_2O_2 digest for the analysis of *Pinus radiata* needles. Commun. Soil Sci. Plant Anal. J. 11:175-188.

27. Schmitt, T. J., M. G. Dosskey, and K. D. Hoagland. 1999. Filter strip performance and processes for different vegetation, widths, and contaminants. J. Environ. Qual. 28:1479-1489.
28. Haag, K. H., R. L. Miller, L. A. Bradner, and D. S. McCulloch. 1996. Water-quality assessment of southern Florida: an overview of available information on surface- and ground-water quality and ecology. U.S. Geol. Survey Water-Resources Invest. Report 96-4177, 42 pp.
29. Cisar, J. L., R.J. Hull, D. T. Duff, and A. J. Gold. 1985. Turfgrass nutrient use efficiency. P. 115. In Agronomy abstracts. ASA, Madison, WI.
30. Titko, S., III, J. R. Street, T. J. Logan. 1987. Volatilization of ammonia from granular and dissolved urea applied to turfgrass. Agron. J. 79:535-540.
31. Bowman, D. C., J. L. Paul, W. B. Davis, and S. H. Nelson. 1982. Reducing ammonia volatilization from Kentucky bluegrass turf by irrigation. Hort. Sci. 22:84-87.
32. Tilman, D., J. Knops, D. Wedin, P. Reich, M. Ritchie, E. Siemann. 1997. The influence of functional diversity and composition on ecosystem processes. Science, 277: 1300-1302.
33. Hooper, D. U., and P. M. Vitousek. 1998. Effects of plant composition and diversity on nutrient cycling. Ecological Monographs, 68:121-149.

Chapter 13

New Tools for the Analysis and Characterization of Slow-Release Fertilizers

J. B. Sartain[1], W. L. Hall, Jr.[2], R. C. Littell[1], and E. W. Hopwood[1]

[1]Soil and Water Science Department, University of Florida, Gainesville, FL 32611
[2]IMC Global Operations, 3095 Country Road, Mulberry, FL 33860

The commercial development of slow-release fertilizer materials has been incremental and based on use of several unique technologies over the last fifty years. While each technology has found a niche in certain specialty markets, none have found widespread use in broad based agricultural markets. Each technology was addressed, as it was developed, in terms of the regulation and analysis of the specific material. Equipped with an effective method to evaluate a broad range of materials instead of a number of product-specific methods, regulators and the fertilizer industry can begin to monitor these materials efficiently in a laboratory setting. The history of the development of a method aimed at accomplishing this goal is described. Key components in the process were development of an accelerated laboratory procedure, development of a stable laboratory soil incubation method mimicking real-life biological conditions, and the correlation of the data from both new methods.

Introduction and Background

The use of fertilizer materials exhibiting characteristics of higher efficiency as compared to reference soluble fertilizers is widespread in certain market segments. These materials have been marketed using many differing names, claims or descriptions for the higher efficiencies of their products. All of these materials have been referred to as a single class of materials called 'Enhanced Efficiency' by the Association of American Plant Food Control Officials (AAPFCO) (1). Within the 'Enhanced Efficiency' class of materials there are two broad categories; **inhibitor** materials and **slow release** materials. These categories will be described as *materials with* urease or nitrification **inhibitors** and **slow release** *materials that delay their nutrient availability for plant uptake relative to a reference soluble material*. Historically nitrogen has been the focus of most of the slow release technology and products developed. The characteristics of nitrogen sources are discussed Carrow, Waddington and Rieke (2).

The commercial development of materials has been incremental and based on use of several unique technologies over the last fifty years. Each technology was addressed, as it was developed, in terms of the regulation and analysis of the specific material. At the time of each development this approach was adequate based on the limited number of products and applications of the materials. However, as the numbers and uses of the products have increased, the individualized approach to regulation has become less and less effective. The widespread use of bulk blending of multiple slow release materials has further complicated the issue and made effective regulation nearly impossible. This is an international issue, not just a North American issue. Trenkel (3) summarized these issues on a worldwide scale. This situation has developed to the point where the checkerboard of regulation and enforcement is inadequate to assure compliance with label claims or delivery of consistent products. There are instances where the system works, but generally regulation is not uniform.

Piecemeal development of technology has also lead to inconsistencies in the analytical methods needed to effectively evaluate the materials, claims, and performance of today's products. Despite the many new technologies of today, there have not been any significant methods of analysis accepted for use (of water insoluble materials) by Association of Analytical Chemists International (AOACI) since

1970 (4). Further, these methods only utilize a two hour time frame and do not standardize temperature. They are also hampered by examining what is not released over time instead of measuring what is released. The analytical methodology needed to measure release is a key to cure the regulatory inconsistencies above. Equipped with a method to evaluate a range of materials, instead of product-specific methods, regulators and producers can monitor materials and blends efficiently in a laboratory. To this end, AAPFCO established a task force to address these regulatory and analytical issues.

Task Force Activities

In 1994, under the joint guidance and authority of AAPFCO and The Fertilizer Institute (TFI), a Controlled Release Fertilizer Task Force was assembled (5) to address issues hindering the effective regulation and analysis of slow release materials. Over the last 8 years the task force has addressed several issues including: a mission statement, terms and definitions proposals, philosophical discussions on the meaning of slow release and other worthwhile topics. However, the overriding issue, and the one occupying the bulk of their time was the development of an effective method to assess slow release in a laboratory setting. This effort was further divided into work on a pure laboratory version based on release acceleration and a soil incubation version that would take into account biological factors. Lab efforts were headed by IMC, soil methods by the University of Florida. Both methods have had many evolutionary twists and turns over the last 8 years. Through the work and input of countless scientists, regulators and other stakeholders, the laboratory method was proposed and collaborated through AOACI in 2000. However, as the data were gathered and additional comments offered, it was evident that there were additional modifications needed. In order to make the method acceptable for routine laboratory use it had to be further automated. The study was withdrawn and another is under design using a more automated procedure. The soil portion of the work was headed by the University of Florida Soil and Water Science Laboratories. The discussion that follows details the specific method development steps and final procedures proposed to AOACI. Having said this, taskforce work will not be complete upon method acceptance. Several issues must be resolved, including: labeling guidelines, how to guarantee release rates, sampling and preparation procedures for unground materials and other regulatory issues.

Laboratory Method Development

As a result of task force input and earlier experience in efforts to measure release of various materials, goals were proposed. These added focus to efforts and helped develop consensus on the direction of the task force. The method development goals are listed in Table I. In addition to these goals, a protocol for collaborating the method was submitted through AOACI. Comments on this protocol also required addressing many additional issues not originally considered in the early development stages. The inputs of the commenters from the fertilizer methods committee are acknowledged in refining the procedure and making it applicable to a wider group of potential users. The collaborators themselves were the best source of real world user information that helped complete the evaluation process. A short scope and synopsis of the method follows to give the reader a background as each development area is discussed.

Table I. Goals of the method development process.

Goal	Purpose
Develop a Structure to Categorize Materials	Organize Materials Into Groups by Technology/Release Mechanism
Status of Current Materials Won't Change	Maintain Status of Materials Currently Making Slow Release Claims
Method Can Be Run in an Analytical Laboratory	Acceptable as a Regulatory Tool Using Standard Laboratory Equipment
Method Can Be Run In Less Than 7 Days	Acceptable as a Regulatory Tool and is Time and Cost Effective
Method Must Be Able To Gain Wide Acceptance	Many States Must Use a Method Before it can be an Effective Regulatory Tool
Would Be Applicable to Both Materials & Blends	Mixtures of Slow Release Materials Pose Problems for Current Methods
Can Be Correlated to Agronomic Data	Release is Claimed Based on Biological Activity - Must Relate to Agronomy
Can Be Used to Extract Multiple Nutrients	N-P-K Also Possibly Secondary and Micronutrients

Principle

A representative unground sample is exposed to increasingly aggressive extractions. Each extraction is designed to isolate nutrients that would become available over time. Each extract is analyzed by AOACI procedures for the nutrient of interest. Along with analysis of total nutrients and method matched reference materials, data are used to develop information specific to the cumulative percent of total nutrient released over time. Samples containing inhibitors should be analyzed according to methodology submitted by their manufacturer; content and purity are determined independently of this procedure. Although present, inhibitors should not alter results of the procedure. This method is applicable to Nitrogen, Phosphate, and Potassium.

Sampling, preparation and sample size. Many of the materials to be tested are granular coated materials that derive their slow release properties from an insoluble or nearly impervious coating. Grinding, or even harsh handling as part of sampling or preparation, is not an option. Although an attempt was made to determine an appropriate weight of sub-sample to analyze, experience and input from others attempting this exercise indicated that a realistic sample size would be limited by factors other than a statistical model. The traditional methodology indicates 1.7 – 2.0 grams of sample be used. This is certainly not enough sample based on a typical granule size of 2-4 mm. The issue is further complicated by the ability to accurately reduce the sample size by use of a riffle or sample splitter. Additionally, equipment and sample solvent ratios also limit the maximum weight of sample. The original sample size used in the first study was 20.0 grams. This size met the sample/solvent ratio, but newer instruments and detection methods (combustion N) perform better at higher N concentrations when analyzing liquid extracts. After due consideration, a 30.0 gram sample size was selected.

Extraction Equipment. After a literature review of numerous methods and protocols, several approaches were developed to optimize the equipment needed to meet the objectives of the task force. There was considerable trial and error in this process, ultimately leading to the conclusion that a jacketed chromatography column proved most effective to perform the test (figure 1). This equipment provided the ability to maintain temperature control and allow a flexible continuous extraction flow scheme. The options needed to include reversal of flows, addition of air bubbles to the column and precise flow rates. A 24 channel, reversible, programmable peristaltic pump was used to move extraction solutions through the system. Because temperature is a key element in all slow release product release characteristics, precise management is accomplished using a water manifold supplied by a centrifugal pump and digitally controlled water bath.

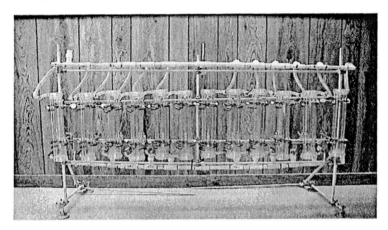

Figure 1. View of lab apparatus using chromatography columns.

Protocol. Given the goals for test duration, an accelerated procedure was needed to rapidly mimic material release usually designed to last several months under typical agronomic conditions. Multiple scenarios were tested; ultimately the final method included use of two extraction solutions (water and 0.2% citric acid), two temperatures 25°C and 65°C, accumulation and combination of extracts, and stabilizing additives. In addition, varied extraction times designed to fit normal work hours (five eight hour shifts per week) were employed to be user friendly while maximizing extraction time.

Detection. The method is designed to produce liquid nutrient extracts to be analyzed by currently accepted methods. Those techniques are already part of the literature and therefore addressed only by reference in the method. However, the ability to analyze liquid extracts containing low to moderate nutrient levels must be verified in order to generate valid data. Consequently, method specific reference materials will be incorporated as part of the final method submission.

Expression of Data. One of the major hurdles of the taskforce has been the issue of how to express/guarantee slow release claims. At the writing of this paper no final determination has been made. However, in order to visually express the results of this method, a graphic form has been established to present results of the method. Traditionally many producers in sales literature and in other scientific presentations have used similar means (6). Below is an example of the graphic presentation of data for a slow release material (sulfur coated urea) SCU (figure 2). The data are graphed as cumulative nutrient extracted over time (in hours) of extraction.

Data generated using the lab method can be used to correlate with the concurrently developed soil incubation method that follows. Keep in mind that although the data generated by the soil method represents release in a controlled agronomic situation, the complex variables of time, temperature, soil, moisture, biological and other conditions make correlation of these data to specific field conditions unlikely. It has been established that the data do represent release of nutrients tested under standardized agronomic conditions. This is important because the information can be used to correlate laboratory data with nutrient release under reproducible biological conditions.

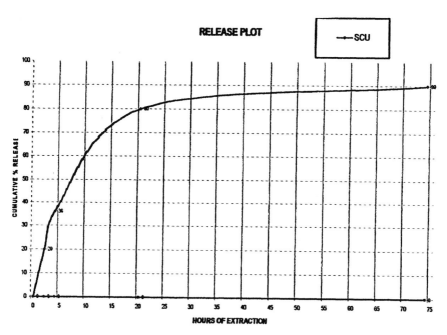

Figure 2. Release plot for SCU illustrating percent released vs. time.

Soil Incubation Methodology Development

Initial Studies: - Plastic Bags. A number of studies were conducted in the development of the currently used methodology. Initially, 250 mg of N from five sources (Milorganite®, SCU, Nitroform®, IBDU® and Urea) was mixed with 250 g of an incubation media. The incubation media was composed of 245 g of a mixture of 85% sand and 15% organic matter (volume basis) and 5 grams of soil. Incubation media containing the desired N source was moistened to approximately field capacity and placed in a plastic bag. These were incubated at room temperature for 7, 14, 28 and 56 days. Triplicate samples were used. After the desired incubation period, the soil was removed from the bags, placed in specifically designed funnels and leached with 200 ml of Deionized (DI) water under vacuum. The leachate was analyzed for nitrate and ammoniacal nitrogen. When the bags were first opened a strong odor of ammonia was observed from some bags. Additionally, leachate color varied according to N source. N Recovery was very poor even for soluble N sources (maximum 60%).

Ball Jar Studies & Ammonia Volatilization. It was believed that the poor recovery was due to the volatilization of N, thus an ammonia trap was placed in the incubation chamber. In this phase of the methodology development, the plastic incubation bags were replaced with incubation jars and a beaker containing 20 ml of 0.2 M sulfuric acid was placed in each jar as an ammonia trap (7). Otherwise, treatments were the same as when plastic bags were used. The solution in the beakers was replaced every 14 days and titrated with base to determine the quantity of N lost through volatilization. In order to determine the efficiency of the N removal by leaching the soils were leached with three 100 ml volumes of DI water and the nitrate and ammoniacal N contents were determined in each. Very little to no N was detected in the third leachate; therefore it appeared that the leaching technique was adequate. However, after accounting for the evolved N only 60 to 80% of the applied N was recovered from the N sources. One of the plausible explanations for the poor N recovery was that it was being fixed in some manner which protected it from leaching.

Organic matter: Leachates from Milorganite, Nitroform, SCU and urea were discolored with a dark brown hue. The IBDU leachate was clear. It was found that the milorganite, nitroform, SCU and urea leachates were alkaline ranging in pH from 8.8 to 9.1, while the IBDU leachate had a pH of 5.6. Based on the color of the leachates, it was assumed that the discoloration was due to soluble organic matter. Additionally, it was suspected that the poor recovery of N from the system was due to fixation of the ammonium N by the organic matter (8). Therefore, in the following studies the incubation media was composed of sand and the inoculant soil.

Based on information gained from the previous studies the following study was designed. The equivalent of 250 mg N from five N sources (ammonium nitrate, SCU, Nitroform, IBDU and Milorganite) was mixed with a media containing 245 g sand and 5 g of an Arredondo fine sand, moisten to 12% moisture (ca. 90% water holding capacity) and incubated at room temperature for 7, 14, 28, 56 and 112 days. A beaker containing 20 ml of 0.2 M sulfuric acid was placed in the incubation jar for ammonia entrapment. Three replicates were used. The beakers were removed and titrated with 0.2m sodium hydroxide every 14 days. After the desired incubation period the soil was leached under vacuum with 300 ml of DI water. These leachates were clear and not discolored. Ammoniacal and nitrate nitrogen were determined on the leachates using a Rapid Flow N Analyzer. At the 7 day sampling in excess of 93% of the N was recovered from the soluble N source.

No volatile ammonia N was detected at the 7 day sampling, however, ammonia N was trapped in all subsequent samplings for some of the evaluated materials. At no time was there ammonia N detected in the ammonium nitrate incubation jars. However, ammonia N represented a significant portion of the N released from the slow-release materials, particularly at the 112 day sampling. As much as 82% of the total N released from SCU was trapped as ammonia N over the 112 day incubation period. Additional methodology modifications were studied to reduce or eliminate the volatile N loss.

Incubation lysimeters: A 30 cm section of 7.5 cm diameter PVC tubing was fitted with a fiberglass mat across the lower end which was held in place by a PVC cap. The cap was drilled and fitted with a barbed plastic fitting and a tygon tube was attached for leachate collection. An additional cap was used on the upper end along with a coating of stop-cock grease to seal the lysimeter. A mixture of an uncoated quartz sand (1710g) and a surface layer (0 to 5 cm depth) of Arredondo fine sand (90g) was mixed with the equivalent of 450mg N from each source and placed in the incubation lysimeters. The same N sources as used in the previous studies were used. The sand/soil/N source mixture was brought to 10% moisture (approximately 80% water hold capacity) by adding 180 ml of 0.01% citric acid. A 50 ml beaker containing 20 ml of 0.2M H_2SO_4 was placed in the head space of the incubation lysimeter as an ammonia trap. The solution in the ammonia trap was replaced and analyzed for NH_4-N by titration every 7 days. After 7, 14, 28, 56, 84, 112, 140 and 180 days each lysimeter was leached with one pore volume of 0.01% citric acid (500 ml) using a vacuum manifold for 2 minutes. Leachate volume was recorded and an aliquot was taken for Urea-N (9), NH_4-N (10) and NO_3-N (11) analysis. Since no volatile N was detected during any of the incubation periods, the three forms of N detected in the leachate were summed for an estimate of the total N released over time.

Nitrification: In all previous incubation methodologies most of the nitrogen was detected as either free ammonia, urea, or ammoniacal nitrogen. Very little nitrate nitrogen was detected. Inclusion of the 0.01% citric acid buffered the pH of the sand/soil/nitrogen source mixture such that highly alkaline conditions and the production of toxic levels of ammonia nitrogen were avoided. As shown in figure 3, most of the nitrogen released from Polyon® after the first 14 days of incubation is in the nitrate form. This suggests that the urea nitrogen released from Polyon® is being nitrified and that the system is microbiologically active thus, ultimately depicting the conditions of a natural soil system (12).

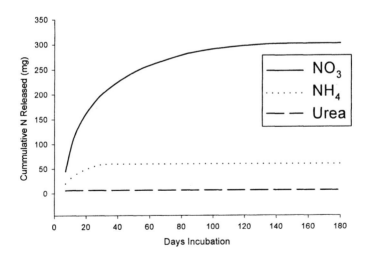

Figure 3. Nitrogen forms detected in lechate from Polyon urea.

Nitrogen Released Over Time: The quantity of nitrogen recovered in the leachate collected at the various sampling intervals was used as an estimate of the release on nitrogen from the source materials over time. Percentage of the nitrogen released over 180 days of incubation for ammonium nitrate, Polyon, SCU, Nutralene, Nitroform, and Milorganite is shown in figure 4. Ninety four percent of the 450 mg nitrogen as ammonium nitrate was accounted for in the first 7 day leachate and by the second leaching (14 days) 96 percent of the nitrogen was recovered.

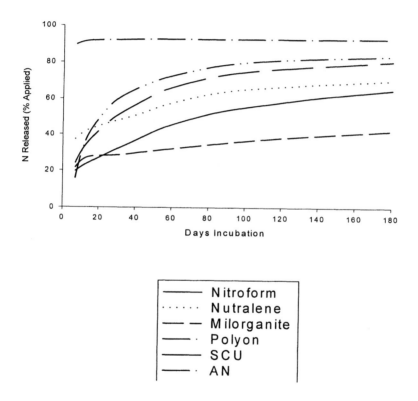

Figure 4. Plot of % release versus days of incubation. for 6 materials.

The slow release materials SCU and Polyon followed similar trends in release resulting in approximately 80% of the applied N released within 112 days. Nutralene and Nitroform also produced similar N release trends with Nutralene consistently releasing a larger quantity of N over time. Approximately 60% of the applied N in Nitroform was released during the 180 day incubation. The N release curve for Milorganite was much flatter than for the rest of the materials with just over 40% of the N released in 180 days.

Approximately 20% of the total N in each of the slow-release materials was recovered in the first 7 day leachate, except for Nutralene which released 37% of the applied N in the first 7 days. By 14 days incubation the rate of N release from the different N sources began to differ significantly. At 84 days incubation, 93, 90, 90, 80 and 82% of the total N released from Polyon, SCU, Nutralene, Nitroform and Milorganite, respectively, had been released. Very small quantities of N were released from all the N sources during the last 70 days of the incubation, except for Nitroform which released an additional 12% of its total N released during this period. Based on these findings (13) incubation periods of greater than 112 days may not be necessary for most slow-release N sources other than some of the methylene ureas.

Correlation of Soil Incubation Nitrogen Release and Accelerated Lab Extraction: Ultimately, the objective of the research is to relate the Accelerated Lab Extraction Procedure to the Soil Incubation Nitrogen Release, such that the Accelerated Lab Extraction Procedure can be used to predict the nitrogen release over time. This relationship was established using non-linear regression techniques. Non-linear regression curves were fitted to the nitrogen release data separately for replication of each material (14).

The functional form of the Soil Incubation Release was found to be

$$\% \text{ Nutrient Released} = a - b * e^{-ct}$$

where a equals the mean value of percent N released when time equals zero, b equals the slope of the function or the mean rate of increase in N released over time, c equals the maximum level of N released or the asymptote, e equals the natural logarithm and t equals time.

As an example, for Polyon this relationship was found to fit the replication 2 and 3 Soil Incubation Nitrogen Release data with a coefficient of determination (R^2) of 0.995 (Figure 5).

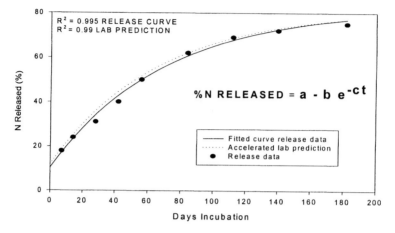

Figure 5. Relationships between soil release and lab prediction data.

Regression analysis was performed on replication 2 and 3 of the data in order to generate the regression parameters and then the Accelerated Lab Extraction data were used to determine how closely the Soil Incubation Release data from replication 1 could be predicted. The parameter values a, b and c were used as dependent variables for replication 2 for each material type and regressed on the Accelerated Lab Extraction data. Predictions of the non-linear regression parameters were computed from the Accelerated Lab Extraction data for all replications. Coefficient of determination for replication 2 for the non-liniar regression of the Accelerated Lab Extraction data were 0.99 (Figure 5). Comparisons of non-linear regression curves with soil replicate 1 indicates how well the Accelerated Lab Extraction data can predict the actual nitrogen release curve when release data are not known. This comparison for Polyon using replication 1 data produces an R^2 of 0.90 (Figure 6). This suggests that the Accelerated Lab Extraction data can predict the nitrogen release from Polyon over time with an accuracy of 90%. This represents a strong relationship considering the data points employed as well as the sampling and analytical variances. At present the relationships are specific to the material being evaluated, but as more data are accumulated it is hoped that a more general relationship can be identified. Based on initial findings at least a grouping of slow release nitrogen sources relative to material type can be achieved.

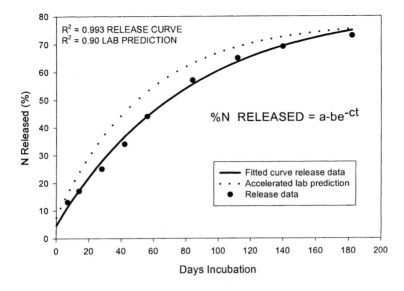

Figure 6. Using lab data to predict soil replicate 1 as an unknown.

Future Work: Several additional Soil Incubation Nitrogen Release and Accelerated Lab Extraction studies need to be run on other slow release nitrogen sources as well as blends of slow release and soluble materials to establish reproducible equations. Through non-linear regression analysis of additional data it will be determined if a single prediction equation can be used for all slow release materials, or if the materials must be grouped according to release characteristics. If a grouping according to type of material is needed it may be necessary to require labeling indicating the percent of materials used in the mixture.

The influence of different soils on nitrification also needs to be investigated before the soils incubation procedure can be recommended in different regions of the country. At present only the influence of an Arredondo fine sand (Grossarenic Paleudults, loamy siliceous, hyperthermic) from Central Florida has been used in the incubation nitrogen release studies. In the future, the influence of soils from different regions of the country will be evaluated.

A weak solution (0.01%) citric acid has been used in the Soil Incubation Nitrogen Release studies to stabilize the medium pH. Concern has been expressed regarding the influence of the citric acid on the 'natural' release of nitrogen from some of the slow release sources. Additional studies are planned to evaluate different concentrations of citric acid on nitrogen release over time and on media pH stabilization.

Conclusions: Based on studies to date it appears that the Accelerated Lab Extraction Procedure can be used to predict the nitrogen release rate of slow release nitrogen sources with acceptable accuracy ($R^2 \geq 0.90$), but at present the procedure is material specific. The Accelerated Lab Extraction Procedure is reproducible, having an average CV of 3.2%. The Soil Incubation Nitrogen Release methodology has a CV of 9.1% and can be expressed in the non linear regression function (% Nitrogen Released = $a - b * c^{-ct}$) with an R^2 of 0.99. Nitrification is occurring due to the predominance of nitrate nitrogen in the leachates of urea source materials which suggests the soil incubation system is being maintained in an aerobic manner and it is microbiologically active.

References:

(1) AAPFCO Publication # 56. 2003. Association of American Plant Food Control Officials, pp 105, D. L. Terry, Secretary, University of Kentucky, 103 Regulatory Services Bldg., Lexington, KY 40546-0275.

(2) R. N. Carrow, D. V. Waddington, P. E. Rieke. 2001. pp 307, Turfgrass Soil Fertility and Chemical Problems, Assessment and Management, Ann Arbor Press, Chelsea, Michigan

(3) Trenkel, M.E. 1997. Improving Fertilizer Use Efficiency – Controlled Release and Stabilized Fertilizers in Agriculture. IFA, Paris.

(4) Official Methods of Analysis of AOAC INTERNATIONAL (2000) 17th Ed., AOAC International, Gaithersburg, MD, USA, **Method 970.04**.

(5) AAPFCO Publication #48. 1995. 164-164. Association of American Plant Food Control Officials, D. L. Terry, Secretary, University of Kentucky, 103 Regulatory Services Bldg., Lexington, KY 40546-0275.

(6) G. L. Smith, Presentation to the AAPFCO Annual Meeting 1995. AAPFCO Publication #48. 1995. 116-120. Association of American Plant Food Control Officials, D. L. Terry, Secretary, University of Kentucky, 103 Regulatory Services Bldg., Lexington, KY 40546-0275.

(7) Ernst, J. W. and H. F. Massey. 1960. The effects of several factors on volatization of ammonia formed from urea in the soil. Soil Sci. Soc. Amer. Proc. 24(2):87-90.

(8) Jensen, H. L. (1965). Nonsymbiotic nitrogen fixation. Agronomy 10:436-480

(9) Mulvaney. R. L. and Bremner, J. M. 1979. A modified diacetyl monoxime method for colorimetric determination of urea in soil extracts. Commun. Soil Sci. Plant Anal. 10:1163-1170.

(10) O'Dell, J. W. 1993a. EPA method 350.1 'Determination of ammonia nitrogen by semi-automated colorimetry. Environmental Monitoring Systems laboratory, Office of Research and Development, U. S. Environmental Protection Agency. Cincinnati, OH.

(11) O'Dell, J. W. 1993b. EPA method 353.2 'Determination of nitrate-nitrite nitrogen by semi-automated colorimetry. Environmental Monitoring Systems laboratory, Office of Research and Development, U. S. Environmental Protection Agency. Cincinnati, OH.

(12) Alexander, M. (1965). Nitrification. Agronomy Journal 10:307-343.

(13) Sartain, J. B. and Kruse, J. K. 2001. Selected fertilizers used in turfgrass fertilization. Soil and Water Sci. Dept. CIR-1262. FL. Coop. Ext. Svc. Univ. Florida. Gainesville, FL.

(14) SAS Institute. 1985. SAS/STAT Guide for personal computers. 6th Ed. SAS Institute, Inc. Cary, NC.

Chapter 14

Impact of High-Yield, Site-Specific Agriculture on Nutrient Efficiency and the Environment

Harold F. Reetz, Jr.

Midwest Director, Potash and Phosphate Institute, 111 East Washington Street, Monticello, IL 61856

Crop production is a complex system of integrated physical, chemical, and biological processes managed under increasing economic and political constraints. Modern production systems employ computer and satellite technology to manage controllable factors on a site-specific basis. Fields are subdivided into management zones, defined by differences in physical and chemical characteristics that can be measured and managed to make more efficient use of available resources. To be competitive, each producer must attempt to optimize---and even maximize---yield. Yet in the aggregate, increased production can depress prices and profitability. Attempts to increase production are often cited as sources of environmental problems, but increasing productivity is a major strategy to protect environmental quality. Site-specific management may provide solutions to these paradoxes.

Integrated, intensive, site-specific systems for crop and soil management are providing a new dimension to crop production. These systems combine traditional management practices and skills with new technologies that help producers meet the challenges of 21st-century farming.

As the technologies and their implementation evolve, the new approaches tend to lose their identity as "new" and become integrated into common practice. Thus it becomes difficult to track the adoption curve after the first few years, and it is nearly impossible to fully assess the economic impact of the new technology on agricultural production and profitability.

Originally used to help identify and manage sources of variability in crop yields, site-specific systems are taking on additional significance as a tool for meeting environmental regulations. As nutrient management planning is integrated into the guidance for developing nutrient management plans for concentrated animal feeding operations (CAFOs) with more that 1,000 animal units, the tools of site-specific management will become increasingly important for sustaining crop production systems. The requirement for plans is broadening to encompass a growing percentage of farms. Eventually some aspects of the nutrient management plan requirement will likely be a part of most farms.

Current research is defining the potential use for these new technologies. The Foundation for Agronomic Research (FAR) and the Potash & Phosphate Institute (PPI) coordinate a multi-state, multi-disciplinary research program to help guide the evaluation of site-specific technology. University researchers and graduate students in 15 states in the Midwest, Mid-South, and Mid-Atlantic regions are cooperating in predominantly on-farm research projects, along with producers, input suppliers, and crop consultants --- over 100 total cooperators. Information on this project is available on the Internet at www.farmresearch.com.

Site-Specific Management Systems

One of the fundamental principles guiding the implementation of site-specific systems for potassium (K) and phosphorus (P) is to increase inputs so that the system is operating above the range of inputs where a further increase in yield is expected (Figure 1.). When managing at a high level of input, reducing that input has less impact on profit potential than when managing at lower input levels. Managing at the higher level provides flexibility in inputs for any given growing season. When input rates are in the lower, more crop-responsive range, there is little flexibility. Input rates cannot be reduced, even for one season, without substantial risk of yield losses.

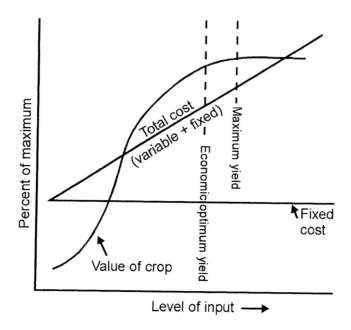

Figure 1. Relationship between total cost, value of crop and level of input, demonstrating that there is more flexibility relative to input decisions when managing at higher input levels than when managing at lower input levels.

Increasing yield goals raises the question of impact on grain prices. The individual producer has little impact, and little control, on the macro-economic forces that set grain prices. But he does have significant impact on his potential yield levels. He can improve soil and crop management practices, increase inputs, and selectively eliminate less-productive parts of fields, all in an attempt to raise average yield levels. All of these decisions can be aided by the use of site-specific tools to focus on smaller areas within fields that may benefit from being managed differently compared to the field average.

A 4-year summary of top profit producers in the Iowa Soybean Association's producer profitability survey, found that 67 percent of the producers in the top profit group produce higher yields. Enhanced marketing skills and reduced input costs were important for 12% and 21%, respectively. Summaries for Kansas and Minnesota showed similar results. These surveys further found that the top profit producers were faster to adopt technology and spent more time gathering information, analyzing choices, planning activities and evaluating results.

Grid sampling, a tool often associated with site-specific management systems, helps identify areas within fields where soil tests vary significantly from the field average. In a 90-acre (36.5 ha) central Illinois field (Figure 2), the average soil K test of 179 ppm, indicated no potash fertilizer was needed. The recommended soil test goal for this field was 175 ppm. But a 1-acre grid soil sampling plan revealed that over half of the field (47 acres) required K fertilizer to reach optimum soil levels. Another 30 acres needed maintenance applications only. The remaining 13 acres had a high soil test and needed no K. The high-testing areas masked the need for K in the rest of the field when all samples were averaged together, while over half of the field was below optimum soil test levels under field-average nutrient management. This is missed market opportunity for the fertilizer dealer and missed production and profit for the producer.

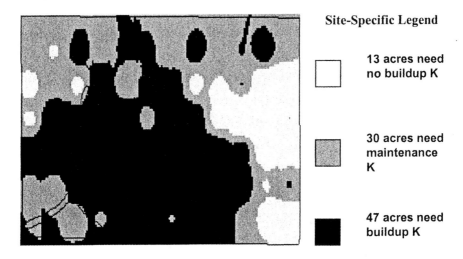

Figure 2. Soil test K recommendations map for a 90-acre (36.5-ha) central Illinois field as delineated by one-acre grid sampling. Soil test K = 179 ppm; Soil test goal = 175. Field average recommendation was no K needed.

Grid size is a subject of debate, but comparison of different sampling plans for a 640-acre Illinois field (Figure 3) shows that field average management would miss 38% of the P fertilizer required for optimum production. Even a course grid sampling plan (330-foot grid) would only miss about 9% of the P requirement (Reetz, 2000). The University of Illinois recommends a 2 ½-acre sampling plan, which this comparison shows would identify 95.5% of the P requirement. Regardless of the position taken on size, it is important to note any common-sized sampling grid is better than application using field-average tests.

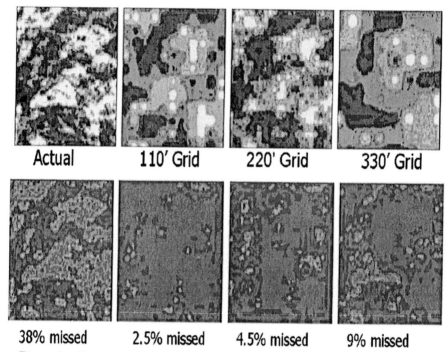

Figure 3. Comparison of P requirement missed by different soil sampling grid sizes for a 640-acre tract in Illinois. Based on simulated distribution of over 1 million soil samples. Applying constant field-average rate under-fertilized 38% needing P. 110-foot grid missed 2.5% needing P; 220-foot grid missed 4.5%; 330-foot grid missed 9%. Course grid was still superior to field average.

Where more information is available, it may be better to use a zone sampling plan. Again there are a variety of ways to establish the pattern to be used, but generally soil survey, past yield maps, and other known patterns of variability are used to establish sampling zones. If zones are larger than the desired sampling density, they can be further subdivided. In any case, using GPS to delineate boundaries of zones is helpful. The GPS coordinates of each sample (or probe site) should be documented to assist in locating spots for further data collection or observation and for future repeat sampling. In an Indiana field, which under field-average management required no buildup potassium, zone sampling identified a need for 10 tons (0.9 mT) of fertilizer K.

Using site-specific management, separate recommendations can be developed for each management zone, and fertilizer can then be applied accordingly using variable-rate application equipment. Producers can track progress in building soil tests over time by repeated soil testing guided by the same GPS coordinates. Through GIS mapping of the test results, difference maps can be generated, and other computations such as nutrient removal maps based on yield, and various economic comparisons can be made on a geo-spatial basis. These tools offer new dimensions to record keeping and decision making.

The value of GIS and related tools can be enhanced by using them with other databases and models that help analyze the data. Current research is employing system dynamics models to help interpret these data and offer new approaches to guiding the nutrient management planning. Having a well-developed database for each field will enhance site-specific management. The database becomes a part of the assets for the field, increasing its potential value. Farms with a detailed, geographically-referenced set of records command a higher cash rent and have enhanced purchase value if put up for sale.

An economic analysis of site-specific management systems is a complex problem. Simple partial budgeting of costs/returns for the technology will often, at best, show a breakeven scenario. It is difficult to put a value on the improved knowledge and decision-making gained for the producer, the landowner, and their advisers by developing and analyzing databases for each field. The value of information can only be realized if action is taken to use that information. This step is not trivial. Many producers who have the information, will for one reason or another, be unable to implement changes to their management.

In other cases there are definite economic advantages that can be readily documented. For example, measuring the yield impact of a wet area of a field can provide a clear estimate of the potential return on investment in an improved drainage system. Variable-rate application of lime often increases profit in site-specific farming. Identifying areas of high or low pH and targeting lime application to specific areas of the field not only provides more efficient use of lime, but also results in more efficient use of other nutrients and more effective use of pesticides, resulting in reduced potential for environmental contamination by agricultural chemicals.

On-Farm Research

Perhaps one of the most significant impacts of site-specific technologies is their use by producers in conducting on-farm research, without the need of specialized plot equipment. Comparisons can be established on a computer and instructions for changes in rates for different product comparisons can be transferred directly to electronic controllers that automatically adjust rates of seed, fertilizer, pesticides or other inputs. Similarly, yield maps and other data collected during the season or at harvest can be stored in a GIS and analyzed for treatment effects by comparing the application map to the yield map. Experimental designs can be developed to utilize statistical analyses that are as accurate as any small-plot field trial, and provides the advantage of using the producers management system, land, and equipment.

While on-farm research trials will not replace small-plot research, and are not appropriate for some kinds of research, they offer new opportunities for rapid adoption of new practices, genetic materials, and other products. By working with producers on a series of on-farm trials, industry and university researchers can "sample" a larger number of environments and geographic locations. Such trials are often useful for field days and demonstrations for other producers in a local area, thus providing an excellent teaching tool. Some new software tools, such as the Enhanced Farm Research Analyst (EFRA) , provide excellent support for on-farm research trial design, implementation and analysis. EFRA is an ArcView "add-on" that was developed in cooperation with the FAR-USB project described earlier, and is available on the Internet for free downloading at www.farmresearch.com. A detailed tutorial package is included.

Efficiency and Environmental Impact

Economic pressures and environmental concerns over the past 20 years have resulted in a reduction in fertilizer use. During the same period, crop yields, and thus nutrient removal from the soil, have steadily increased. The net effect is that soil test levels are declining, in some areas reaching critical levels at which crop production is being limited.

Soil Test Levels in North America, a 2001 summary of soil tests from soil testing laboratories throughout North America shows that approximately half of the samples analyzed are below optimum, that is they fall within the test levels expected to have an economic response to P and K fertilizer (Figure 4). It is suspected that a large percentage of the responsive samples come from fields that also have enough samples testing above the responsive level, so that when recommendations are made for field-average fertilization, they overshadow the higher tests and insufficient fertilizer is applied to correct the low-testing areas.

Increased incidence of visible potassium deficiency symptoms in Midwest corn and soybean fields in recent years is an indication that the cutback on fertilizer use is limiting yields. Economic yield loss occurs long before deficiency symptoms are visible. Again, site-specific management should help identify these potential deficient areas and guide corrective action before economic losses get too high.

Establishing proper soil fertility levels is critical for optimum yields, efficient use of inputs. For example, maintaining a high soil test K level ensures that more nitrogen (N) fertilizer is used by the crop and less remains in the soil after harvest (Figure 5). With high K soil tests, the optimum yield was higher, and it was achieved at a lower N rate (Johnson, et al, 1997; Murrell and Munson, 1999). Similar benefits have been shown for interactions of high phosphorus soil tests with K and N.

(a)

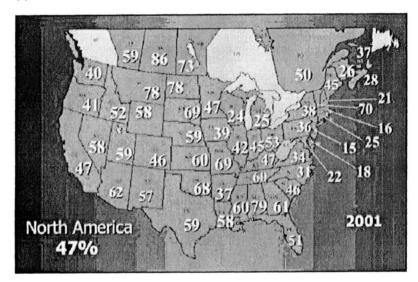

(b)

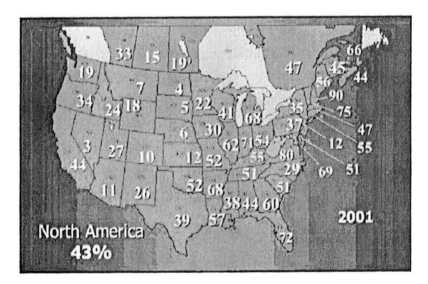

Figure 4. Soil Test Summary for North America, 2001. (a) % of samples testing medium of below in P; (b) % of samples testing medium or below in K. Summary of 2.5 million samples from commercial and university labs by the Potash & Phosphate Institute.

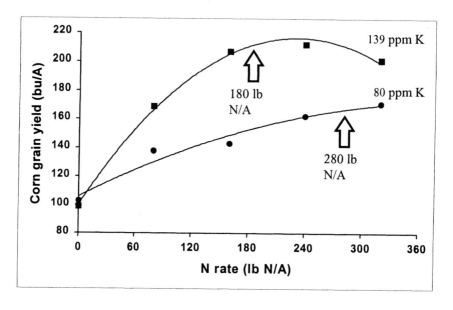

Figure 5. Influence of soil test K levels on corn yields (averaged over 4 years) illustrating rate of N fertilizer necessary to achieve optimum yields with proper soil test K. (Johnson et al., 1997).

The economics of P and K fertilization are such that there is rarely an economic justification for not maintaining soil tests at the optimum level. Identifying such situations and applying fertilizer to correct the deficiency is one of the strongest selling points for site-specific management. In such cases, site-specific management supports improved yields, better nutrient efficiency and reduced environmental degradation.

Conclusions

High yield management helps lower cost per unit of production by spreading fixed costs (land, labor, equipment, etc.) over more units. It requires forward planning, good record keeping, and most important, making a commitment to work toward better management. The tools of site-specific systems (satellites, computers, etc.) are not the goal, but rather the vehicle to help facilitate reaching the goal of improved productivity and profit.

In most of the world, the primary limiting factor to increased production is the lack of adequate land resources. Due to increasing world population and conversion of farmland to other uses, the per capita land resource is steadily declining. U.S. farmers are blessed with the best combination of highly productive soil and favorable climate in the world. We need to protect, and also efficiently utilize those resources to maintain our advantage as a world leader in food and fiber production. As land is shifted to more intensive production in other parts of the world to meet local food needs, fragile ecosystems are upset and wildlife habitat is destroyed. The most productive land is already in production, but higher yields can improve the efficiency with which it is used.

Increasing production of the more productive soils, such as in the primary grain production regions of the U.S can slow that trend. Managing for higher yields also produces more biomass that helps control erosion, improve water use efficiency, and capture more nutrients in crop residue for slow release to subsequent crops.

The world record corn yield of 370 bu/A (23.2 metric tons/ha) set by central Illinois farmer, Herman Warsaw, in 1985, was not surpassed until 1999, when northeast Iowa farmer, Francis Childs produced 394 bu/A (24.7 metric tons/ha), which he eclipsed in 2001 with a 408 bu/A (25.6 metric tons/ha) yield. These and other top yield producers pay attention to details, constantly looking for the next limiting factor in their best fields. Progress is made one step at a time, maintaining high levels of inputs, careful attention to soil conditions, and systematically identifying and eliminating the next yield-limiting factor. The gap between the record yields and the average farmer yield represent the potential for increasing farm productivity, increasing fertilizer markets, and ultimately meeting the world's food and fiber needs while protecting precious soil and water resources.

By focusing on the more productive fields and keeping less productive, more vulnerable fields out of production, more land can be made available for conservation uses, recreation, and other uses. Higher yields, and more profitable farmers also help to improve the business opportunities and general well-being of rural communities.

Agriculture in the 1960s was focused in crop and soil management for improving yields. In the 1970s, attention turned to pest management with the advent of a wide selection of pesticide options along with cultural controls. Marketing became the main thrust of the 1980s as new world markets opened up. In the 1990s, resource management was the leading topic, looking for ways to protect the limited resources available. As the 21st century begins, management of data and information appears to be the next new focus of attention. None of the other focal points has gone away, but at center stage is how information gathered for management decisions is analyzed and interpreted for better-informed decisions.

REFERENCES

Enhanced Farm Research Analyst (EFRA) See Internet website: http://www.farmresearch.com/efra/

Johnson, J. W., T. S. Murrell, and H. F. Reetz, Jr. 1997. "Balanced Fertility Management: A Key to Nutrient Use Efficiency". Better Crops 81(2): 3-5.

Murrell, T. S., and R. D. Munson. 1999. "Phosphorus and Potassium Economics in Crop Production: Putting the Pieces Together". Better Crops 83(3): 28-31.

Reetz, Harold F., Jr. 2000. "Map Making for Variable Rate Fertilization." Better Crops 84(2): 18-20.

Reetz, H. F., Jr., T.S. Murrell, and L.J. Murrell. 2001. "Site-Specific Nutrient Management: Production Examples". BETTER CROPS 85(1): 12-17.

Soil Test Levels in North America. 2001. PPI/PPIC/FAR Technical Bulletin 2001-1, Potash & Phosphate Institute, Norcross, GA 30092.

Rainfall Simulators

One method to help study factors affecting nutrient losses from agriculture is the rainfall simulator. Rainfall simulators reproduce conditions consistent with storm events that generate water runoff. Because of numerous advantages, the use of rainfall simulators for scientific studies of erosion and infiltration have been utilized for many years (*6, 7, 8*). The scientific validity of using rainfall simulation for research work to evaluate nutrient losses in runoff has also been well documented (*9, 10, 11, 12, 13, 14*).

Several different types of rainfall simulators have been described in the scientific literature, including simulators used in laboratory studies (*15*), as well as both microplot and mesoplot scale setups (*16*) used in field research. For example, Sharpley and Moyer (*15*) described a rainfall simulator used to produce rain for soil columns in a laboratory study, while Sumner et al. (*16*) described a rainfall simulator for mesoplot runoff studies, in which rainfall was simulated in a 600 m^2 area. Most rainfall simulators use the microplot size study area and have been used for a variety of scientific studies of non-point nutrient losses from agriculture (*12*). For example, in the studies described by Torbert et al. (*17, 18*) a rainfall simulator was used similar to that described by Miller (*19*), using a Spraying Systems Wide Square Spray 30 WSQ nozzle at a nominal rate of 125 mm h^{-1}, producing a drop size of 2.5 mm, and kinetic energy of 23 J m^{-2} mm^{-1}. In that case, an 1-m^2 area on 2-3% slope was used as the plot size by surrounding the plot with a metal frame driven 0.1 m into the soil to define the study area. Rainfall application was also made to a 10-m^2 region around the study site which was surrounded by a tarp curtain to prevent interference of drops by wind. Similar systems were used to collect most of the data described here.

Conservation Tillage

One effective means of reducing non-point source pollution from crop land is the use of conservation tillage systems. These systems are known to be very effective in reducing erosion and limiting the amount of nutrients that leave the field in sediment (*20, 21, 22, 23, 24, 25*). Potter et al.(*8*) found that maintaining a residue cover on a heavy clay soil in no-tillage systems preserved infiltration rates and controlled sediment losses in erosion. The sediment component of runoff generally has been shown to carry most of the plant nutrients off the field (*26, 11, 27*).

In clay soils of the Texas Blackland Prairie, Torbert et al.(*17*) reported that chisel-tillage had higher total P losses in runoff than no-tillage. In that case, P losses of 0.05 and 0.21 kg P ha^{-1} for no-tillage compared to 0.75 and 1.0 kg P ha^{-1} for chisel-tillage, were observed from rainfall simulation under dry and wet soil conditions, respectively (Table I). The chisel-tillage resulted in greater total P losses because of greater runoff losses of sediment compared to no-tillage. In another study of heavy clay soils, Torbert et al. (*18*) reported that total mean sediment lost during a 3 h simulated rainfall event was significantly reduced in conservation tillage (0.03 Mg ha^{-1}) compared with conventional tillage (0.67 Mg ha^{-1}), which resulted in a 12-fold increase in nutrient losses associated with sediment (Table II). Erosion control is especially important in no-tillage systems because stratification of nutrients can occur near the surface without tillage mixing the surface soil. Potter and Chichester (*28*) found that in the soil surface (0 to 25 mm) P concentrations were about 50% greater after 10 years of continuous no-tillage compared to annually tilled soils.

While the nutrient concentration in the sediment portion of runoff is greatly reduced with surface residue cover, several studies have shown that the concentration of PO$_4$-P in the solution phase is often increased with conservation tillage (*21, 29, 24, 30, 17*). For example, studies in the Blackland Prairie soils of Texas (*17*) indicated that the P content in runoff was increased with no-tillage (0.14) compared to chisel-tillage (0.03 kg PO$_4$-P ha^{-1}). With chisel-till, the PO$_4$-P content in runoff was relatively constant, varying between 0.004 and 0.008 kg PO$_4$-P ha^{-1} during the 30 min runoff event (Fig. 1). In contrast, no-tillage PO$_4$-P content in runoff continued to increase for the entire 30 min runoff event, from 0.004 kg PO$_4$-P ha^{-1} at 5 min to 0.041 kg PO$_4$-P ha^{-1} at 30 min.

This increase in soluble P in runoff under conservation tillage has been attributed to the lack of incorporation of fertilizers into the surface layer (*31, 32, 33*). Timmons et al. (*31*) reported that nutrient losses declined as the level of fertilizer incorporation increased. Other likely causes of this effect may be nutrient stratification near the soil surface (*28*), or the decomposition of plant materials on the soil surface (*34, 35*). Several studies have attributed increased soluble PO$_4$-P in runoff to leaching of P from plant material exposed to rainfall (*29, 36, 37*). However, Chichester and Richardson (*38*) reported that while not significant, means for soluble P losses were higher in conventional tillage compared to no-tillage when measured from watersheds of a Vertisol over a year period. This indicates that the relative importance of nutrient leaching from plant residue may diminish with time.

Fertilizer Application

While conservation tillage practices can result in many environmental benefits, fertilizer application in these systems can be difficult, because of the need to limit disturbance of surface residues that provide erosion control. Several studies have reported higher nutrient losses in runoff when fertilizer was applied to the soil surface as compared to subsurface fertilizer application (*39, 21, 31, 32*). Beyrouty et al. (*39*) reported a 20-40% increase in fertilizer recovery at the end of the year when urea-ammonium nitrate (UAN) solution was applied subsurface compared to

Table I. Influence of tillage system on runoff nutrient losses during rainfall simulation[1].

Tillage	Sediment	Sediment P	Sediment N	Fertilizer N
	(kg ha^{-1})			
	Dry Run[2]			
No-tillage	40a	0.05a	0.12a	0.002 a
Chisel-tillage	1228b	0.75b	1.51b	0.015 b
	Wet Run			
No-tillage	83a	0.21a	0.57a	0.009 a
Chisel-tillage	1699b	1.04b	2.09b	0.031b

[1]Values represent means of 4 replicates. Values followed by the same letter do not differ significantly (0.10 level).
[2]Rainfall simulation was conducted under relatively dry soil moisture condition "dry run" (10%) and relative wet soil moisture condition "wet run"(45%).

Table II. Influence of surface residues and soil moisture conditions on losses of sediment and total N[1].

Management	Dry Run[2]		Wet Run	
	Sediment (Mg ha^{-1})	Total N (kg ha^{-1})	Sediment (Mg ha^{-1})	Total N (kg ha^{-1})
No-tillage	0.01 a	0.02 a	0.03 a	0.07 a
Chisel-tillage	0.25 b	0.62 b	0.67 b	1.29 b
Sod	0.01 a	0.05 a	0.01 a	0.07 a

[1]Values represent means of 3 replicates. Values followed by the same letter are not significantly different ($\alpha = 0.10$ level).
[2]Rainfall simulation was conducted under relatively dry soil moisture condition "dry run" (35%) and relative wet soil moisture condition "wet run"(50%).

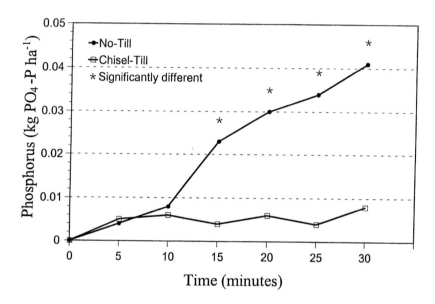

Figure 1. Effect of tillage system on PO_4-P content of runoff solution during a 30 minute runoff event.

surface application. Timmons et al. (*31*) reported nutrient losses decreased with increased level of incorporation of applied fertilizers. However, the losses of fertilizer due to incorporation may be very soil-type dependent. Recent research on a Vertisol found little or no significant differences in agronomic response (grain yield and fertilizer efficiency as measured in grain yield) to eight different fertilizer application methods (*40*). In a rainfall simulation study, Torbert et al. (*17*) also found that very little N in runoff could be attributed to liquid fertilizer applied in a surface band to dry soil. Using ^{15}N techniques to trace the fertilizer, they found that only an average loss of 1.6 kg N ha^{-1} in runoff during a 30 min rainfall event could be attributed to the application of 135 kg fertilizer N ha^{-1}.

Recently fertilizer application equipment has been developed that allows for subsurface application of fertilizers with minimal surface residue disturbance. For example, a spoke wheel applicator applies fertilizer solution with a point injection below the soil surface (*41*) and a coulter-nozzle apparatus that shoots a solid stream of liquid fertilizer into a slit opened behind a rolling coulter (*42*). In the latter study, the fertilizer P was concentrated in the surface 0 to 40 mm in a concentrated band. Morrison and Chichester (*43*) examined several fertilizer applicator designs and found that the coulter-nozzle fertilizer applicator made the least soil disturbance.

Soil Moisture and Tillage System

While the incorporation of fertilizers have been shown to reduce nutrient losses, because of other agronomic and economic considerations, such as product and equipment availability, application of dry fertilizers to the soil surface is likely to continue. As mentioned, subsurface applications in conservation tillage systems are difficult because of the need to limit soil disturbance and subsurface application in pasture is rare due to the resulting damage to the grass. While the application of fertilizer to the soil surface will likely continue, the environmental impact of surface application of fertilizer may be reduced with wise application timing.

Soil moisture condition can be an important factor to consider in the timing of fertilizer application. Torbert et al. (*18*) examined fertilizer application as affected by soil moisture and soil tillage system. Rainfall simulations were conducted on three different surface residue conditions: chisel-tillage, conservation tillage, and bermudagrass [*Cynodon dactylon* (L.)] sod. Rainfall was simulated under relatively dry soil (35% soil moisture) (dry run) and relatively wet soil (50% soil moisture) (wet run) conditions. Rain was initiated under antecedent dry conditions and continued for 3 h, resulting in the wet soil condition. After 48 h, simulated rainfall was applied to the wet soil and continued for another 3 h. Granular fertilizer application was made to separate plots to both dry soil and wet soil in each of the three surface residue treatments immediately before rainfall simulation. A second rainfall simulation (wet run) was conducted for the plot receiving fertilizer application on dry soil. Rainfall simulation was also performed with no fertilizer application (control) under both wet and dry soil conditions. The fertilizer applications to the runoff plots were made as granular 16-20-0, which is a mixture of 42% monoammonium phosphate ($NH_4H_2PO_4$) and 58% ammonium sulfate (($NH_4)_2SO_4$) at a rate which provided 134 kg N ha^{-1} and 168 kg P_2O_5 ha^{-1} (74 kg P ha^{-1}).

Consistent with previous research concerning runoff losses of sediment, sediment and total N losses in runoff during rainfall simulations were greatly reduced with treatments that included surface residues (i.e., no-tillage and sod) (Table II). These reductions were noted for both the dry and wet runs. Runoff losses were greatly affected by both the surface residue and the fertilizer application timing with regard to soil moisture.

The concentrations (mg L^{-1}) and loads (kg ha^{-1}) of PO_4-P in runoff solution are illustrated in Figs. 2 and 3, respectively. These data demonstrate the influence of surface residue management and soil moisture effect on runoff losses of nutrients for heavy clay soils. Overall, the loss patterns of PO_4-P concentrations (Fig. 2) during rainfall simulation were not greatly different from the patterns observed for PO_4-P loads (Fig. 3). This indicated that the runoff nutrient losses were dominated by PO_4-P concentration and not by the total volume of runoff water. This implies that factors that impact nutrient concentration will be the most important aspect determining losses of fertilizer from clay soils.

Nutrient losses in runoff were much larger when fertilizer applications were made to wet soil compared with when fertilizer was applied to dry soil. In fact, when fertilizer was applied to dry soil, combined nutrient losses from both the wet and dry soil rainfall simulations were less than the losses that occurred when fertilizer was applied to the wet soil (Table III).

This study indicated that the time to runoff initiation may be a major mechanism that determines fertilizer concentration in runoff. Granular fertilizer applied to the soil surface must dissolve before entering the soil during water infiltration. It is important to note that most of the nutrient loss during the 3 h simulation in all surface residue treatments occurred within the first 40 min of runoff initiation (Figs. 2 and 3). Any mechanism that either increases the rate of water infiltration or delays the initiation of runoff, increases the amount of fertilizer that moves into the soil and thus minimizes immediate loss in runoff water.

The cumulative amounts of PO_4-P lost in solution for the rainfall simulations are presented in Table III. With fertilizer applied to wet soil, no-tillage reduced the cumulative loss of PO_4-P nearly 7-fold compared with chisel-tillage. This reduction resulted from both increases in time before the initiation of runoff and lower nutrient concentrations once runoff was initiated (Figs. 2 and 3).

The largest cumulative loss of nutrients in runoff occurred with the sod surface residue treatment. This resulted from both a quicker initiation of runoff compared with the tilled treatments and an increase in nutrient concentrations during the

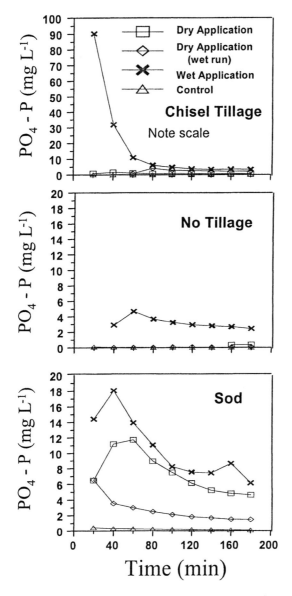

Figure 2. The concentration of PO_4-P in runoff solution during a simulated runoff event as affected by surface residue condition (chisel tillage, no-tillage, or sod) and soil moisture condition for fertilizer application (dry soil or wet soil).

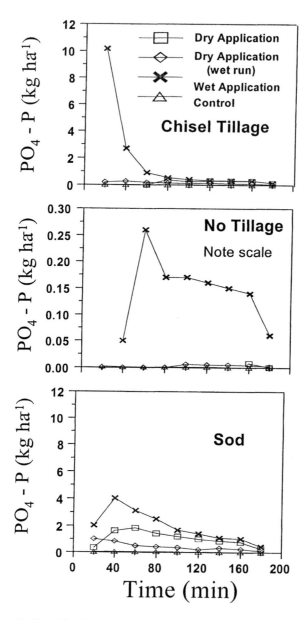

Figure 3. The PO_4-P content (load) in runoff solution during a simulated runoff event as affected by surface residue condition (chisel tillage, no-tillage, or sod) and soil moisture condition for fertilizer application (dry soil or wet soil).

Table III. Influence of surface residues fertilizer application timing on cumulative NH_4^+-N and PO_4^--P losses in runoff [1].

Management	Dry Appl.[2]	Dry Appl. (wet run)	Wet Appl.	Control
NH_4^+-N (kg ha^{-1})				
No-tillage	0.02 A x	0.10 A x	3.88 B x	0.01 A x
Chisel-tillage	3.00 A x	4.82 A x	18.91 B x	0.00 A x
Sod	21.02 A y	6.84 B x	1.92 C y	0.08 D x
PO_4^--P (kg ha^{-1})				
No-tillage	0.01 A x	0.02 A x	1.16 B x	0.01 A x
Chisel-tillage	1.52 A x	1.27 A x	7.96 B y	0.02 A x
Sod	9.27 A y	3.68 B x	17.35 C z	0.25 D x

[1] Values represent means of 3 replicates. Values in the same column followed by the same lower case letter (x,y,z) do not differ significantly (α= 0.10 level). Values in the same row followed by the same upper case letter (A, B, C, D) do not differ significantly (α= 0.10 level).

[2] Dry Appl. indicates that fertilizer was applied under relatively dry soil moisture conditions; Wet Appl. indicates that fertilizer was applied under relatively wet soil moisture conditions.

runoff events. The nutrient loss in runoff during the wet run with fertilizer applied under dry conditions remained high for sod, unlike the chisel-tillage and no-tillage that had nutrient concentrations only slightly above that measured with the control (Figs. 2 and 3). With the sod, approximately 41% of the PO_4-P fertilizer applied was lost in the cumulative runoff when fertilizer was applied to the wet soil, and represented a 25% increase in PO_4-P losses compared with both the rain simulation runs combined when fertilizer was applied to dry soil. This indicated that in a heavy clay soil under wet soil conditions, granular fertilizer application to pastures may result in a significant contribution to runoff loading of surface waterways.

Organic Fertilizers (Manure)

The greatest potential for non-point P contribution to surface waters usually occurs in watersheds with intensive animal production (44). Manure collected from concentrated animal feeding operations (CAFO) has traditionally been applied to fields near the operation because this is a practical means of both improving soil physical conditions and providing needed plant nutrients for crop production. However, long-term manure application to soils at rates exceeding crop uptake can result in elevated soil P levels (45, 46, 47). The N:P ratio of animal manure ranges from 2:1 to 8:1, depending on animal species (48). This N:P ratio is much narrower than required for crop production, resulting in an over application of P and a buildup of soil P levels over time (49).

Research has shown that high loading of P in excess of plant needs directly influences the amount of P found in soil and runoff (50, 49, 51, 13, 14). Research by Edwards et al. (52) has shown that the contribution of P from soils with elevated soil test P is potentially more important and difficult to manage than improper land application of animal manure. This study found that soils with elevated soil test P levels were responsible for 65 to 90% of annual P loss from the watershed even when a major surface runoff event occurred 1 day after manure application to a fescue (*Fescue arundinacea* Schreb) pasture.

Soil Test P and Runoff P

Research has shown that soil P level is directly related to runoff losses of P (50, 49, 51, 13, 14, 53). However, work by Pote et al. (54) and Sharpley et al. (47) demonstrated that the relationship between soil P level and runoff P varied markedly by soil type. Other research by Torbert and Daniel (55) indicated that while a significant relationship between soil test P and runoff P existed, there were large differences in the potential level of P runoff losses at the same soil test P concentration across different soil types within the same watershed. In this case, the relationship between soil P level and runoff P losses were shown to be reduced in calcareous soils compared to non-calcareous soils. At the highest soil test P level (360 mg kg^{-1} soil), maximum concentrations of 1.73 and 1.63 mg L^{-1} P were observed in runoff from two non-calcareous soils, compared to 0.74 and 0.47 mg L^{-1} P in runoff from the two calcareous soils, respectively.

Pote et al. (*54*) indicated that soil physical effects (rainfall infiltration rates) could be useful in equating differences between soil types for losses of runoff P in relation to soil test P. In this work, the difference in predictive equations between soil types were virtually eliminated by accounting for differences between runoff levels. Results from such rainfall simulation studies indicate that a soil test for environmental P can be developed, but it will probably require establishing different criteria for soil test P levels for different soils or classes of soils. For soil test P to be used as a management tool in land application of manure, soils will need to be grouped into reasonable management categories and reliable predictive equations for potential P loss will need to be developed for these soil categories. Such predictive equations, if properly described, could be utilized with tools such as the P index of Lemunyon and Gilbert (*56*) for manure management.

References

1. Rabalais, N.N.; Turner, R.E.; Wiseman, W.J., Jr. Hypoxia in the Gulf of Mexico. *J. Environ. Qual.* **2001**, 30, 320-329.
2. Goolsby, D.A.; Battaglin, W.A.; Aulenbach, B.T.; Hooper, R.P. Nitrogen input to the Gulf of Mexico. *J. Environ. Qual.* **2001**, 30, 329-336.
3. Parry, R. Agricultural phosphorus and water quality: A U.S. Environmental Protection Agency perspective. *J. Environ. Qual.* **1998**, 27, 258-261.
4. Daniel, T.C.; Sharpley, A.N.; Lemunyon, J.L. Agricultural phosphorus and eutrophication: A symposium overview. *J. Environ. Qual.* **1998**, 27, 251-257.
5. Correll, D.L. The role of phosphorus in the eutrophication of receiving waters: A review. *J. Environ. Qual.* **1998**, 27, 261-266.
6. Bubenzer, G.D.; Meyer, L.D. Simulation of rainfall and soils for laboratory research. *Trans ASAE* **1965**, 6, 73-75.
7. Moore, I.D.; Hirschi, M.C.; Barfield, B.J. Kentucky rainfall simulator. *Trans. ASAE* **1983**, 24, 1085-1089.
8. Potter, K.N.; Torbert, H.A.; Morrison, J.E. Jr. Management effects on infiltration, runoff, and sediment losses on Vertisols. *Trans. ASAE* **1995**, 38, 1413-1419.
9. Meyer, L.D. Simulation of rainfall for soil erosion research. *Trans. ASAE* **1965**, 8, 63-65.
10. Niebling, W.H.; Foster, G.R.; Nattermann, R.A.; Nowlin, D.; Holbert, P.V. Laboratory and field testing of a programmable plot sized rainfall simulator.

Proc. Symp. on Erosion and Sediment Transport Measurement, Florence, Italy. 1981, Int. Assoc. Hydrologic Sci. Pub. No. 133.
11. Andraski, B.J.; Mueller, D.H.; Daniel, T.C. Phosphorus losses in runoff as affected by tillage. *Soil Sci. Soc. Am. J.* **1985**, 49, 1523-1527.
12. Edwards, D.R.; Norton, L.D.; Daniel, T.C.; Walker, J.T.; Ferguson, D.L.; Dwyer, G.A.. Performance of a rainfall simulator. *Arkansas Farm Res.* **1992**, 41, 13-14.
13. Sharpley, A.N. Dependence of runoff phosphorus on extractable soil phosphorus. *J. Environ. Qual.* **1995**, 24, 920-926.
14. Pote, D.H.; Daniel, T.C.; Sharpley, A.N.; Moore, P.A. Jr., Edwards, D.R.; Nichols, D.J. Relating extractable soil phosphorus to phosphorus losses in runoff. *Soil Sci. Soc. Am. J.* **1996**, 60, 855-859.
15. Sharpley, A.; Moyer, B. Phosphorus forms in manure and compost and their release during simulated rainfall. *J. Environ. Qual.* **2000**, 29, 1462-1469.
16. Sumner, H.R.; Wauchope, R.D.; Truman, C.C.; Dowler, C.C.; Hook, J.E. Rainfall simulator and plot design for mesoplot runoff studies. *Trans. ASAE* 1996, 39, 125-130.
17. Torbert, H.A.; Potter, K.N.; Morrison, J.E. Jr. Management effects on fertilizer N and P losses in runoff on expansive clay soils. *Trans. ASAE.* **1996**, 39, 161-166.
18. Torbert, H.A.; Potter, K.N.; Hoffman, D.W.; Gerik, T.J.; Richardson, C.W. Surface residue and soil moisture affect fertilizer loss in simulated runoff on a heavy clay soil. *Agron. J.* **1999**, 91, 606-612.
19. Miller, W.P. A solenoid-operated, variable intensity rainfall simulator. *Soil Sci. Soc. Am. J.* **1987**, 51, 832-834.
20. Meyer, L.D.; Wischmeier, W.H.; Foster, G.R. Mulch rate required for erosion control on steep slopes. *Soil Sci. Soc. Am. Proc.* **1970**, 34, 928-931.
21. Römkens, M.J.M.; Nelson, D.W.; Mannering, J.V. Nitrogen and phosphorus composition of surface runoff as affected by tillage method. *J. Environ. Qual.* **1973**, 2, 292-295.
22. Lindstrom, J.J.; Gupta, S.C.; Onstad, C.A.; Larson, W.E.; Holt, R.F. Tillage and crop residue effects on soil erosion in the Corn Belt. *J. Soil Water Conserv.* **1979**, 34, 80-82.
23. Angle, J.S.; McClung, G.; McIntosh, M.S.; Thomas, P.M.; Wolf, D.C. Nutrient losses in runoff from conventional and no-till watersheds. *J. Environ. Qual.* **1984**, 13, 431-435.
24. McDowell, L.L.; McGregor, K.C. Plant nutrient losses in runoff from conservation tillage corn. *Soil Tillage Res.* **1984**, 4, 79-91.
25. Gilley, J.E.; Finker, S.C.; Varvel, G.E. Slope length and surface residue influences on runoff and erosion. *Trans. ASAE* **1987**, 39, 148-152.

26. Barisas, S.G.; Baker, J.L.; Johnson, H.P.; Laflen, J.M. Effects of tillage systems on runoff losses on nutrients, a rainfall simulation study. *Trans. ASAE* **1978**, 21, 893-897.
27. Owens, L.B.; Edwards, W.M. Tillage studies with a corn-soybean rotation: Surface runoff chemistry. *Soil Sci. Soc. Am. J.* **1993**, 57, 1055-1060.
28. Potter, K.N.; Chichester, F.W. Physical and chemical properties of a Vertisol with continuous controlled-traffic, no-tillage management. *Trans. ASAE* **1993**, 36, 95-99.
29. McDowell, L.L.; McGregor, K.C. Nitrogen and phosphorus losses from no-till soybean. *Trans. ASAE.* **1980**, 23, 643-648.
30. Alberts, E.E.; Spomer, R.G. Dissolved nitrogen and phosphorus in runoff from watersheds in conservation and conventional tillage. *J. Soil Water Conserv.* **1985**, 40, 153-157.
31. Timmons, D.R.; Burwell, R.E.; Holt, R.F. Nitrogen and phosphorus losses in surface runoff from agriculture land as influenced by placement of broadcast fertilizer. *Water Resour. Res.* **1973**, 9, 658-667.
32. Whitaker, F.D.; Heineman, H.G.; Burwell, R.E. Fertilize corn adequately with less nitrogen. *J. Soil Water Conserv.* **1978**, 33, 28-32
33. Baker, J.L.; Laflen, J.M. Effects of corn residue and fertilizer management on soluble nutrient runoff losses. *Trans. ASAE* **1982**, 25, 344-348.
34. Johnson, H.P.; Baker, J.L.; Sharader, W.D.; Laflen, J.M. Tillage system effects on sediment and nutrients in runoff from small watersheds. *Trans. ASAE* **1979**, 22, 1110-1114.
35. Mostaghimi, S.; Dillaha, T.A.; Shanholz., V.O. Influence of tillage systems and residue levels on runoff, sediment, and phosphorus losses. *Trans. ASAE* **1988**, 31, 128-132.
36. Schreiber, J.D. Leaching of nitrogen, phosphorus, and organic carbon from wheat straw residues: II. Loading rate. *J. Environ. Qual.* **1985**, 14, 256-260.
37. Schreiber, J.D.; McDowell, L.L. Leaching of nitrogen, phosphorus, and organic carbon from wheat straw residues: I. Rainfall intensity. *J. Environ. Qual.* **1985**, 14, 251-256.
38. Chichester, F.W.; Richardson, C.W. Sediment and nutrient loss from clay soils as affected by tillage. *J. Environ. Qual.* **1992**, 21, 587-590.
39. Beyrouty, C.A.; Nelson, D.W.; Sommers, L.E. Transformations and losses of fertilizer nitrogen on no-till and conventional till soils. *Fert. Res.* **1986**, 10, 135-146.
40. Chichester, F.W.; Morrison, J.E., Jr. Agronomic evaluation of fertilizer placement methods for no-tillage sorghum in Vertisol clays. *J. Prod. Agric.* **1992**, 5, 378-382.
41. Baker, J.L.; Colvin, T.S.; Marley, S.J.; Dawelbeit, M. Improved fertilizer management with a point-injector applicator. ASAE 1985, Pap. 85-1516. ASAE, St. Joseph, MI.

42. Morrison; J.E., Potter, K.N. Fertilizer solution placement with a coulter-nozzle applicator. *Appl. Engr. Agric.* **1994**, 10 (1), 7-11.
43. Morrison, J.E. Jr.; Chichester, F.W. Subsurface fertilizer applicator for conservation-tillage research. *Appl. Engr. Agric.* **1988**, 4(2), 130-134.
44. Sims, J.T.; Edwards, A.C.; Schoumans, O.F.; Simard, R.R. Integrating soil phosphorus testing into environmentally based agricultural management practices. *J. Environ. Qual.* **2000**, 29, 60-71.
45. Sims, J.T. Environmental soil testing for phosphorus. *J. Prod. Agric.* **1993**, 6, 501-507.
46. Snyder, C.S.; Chapman, S.L.; Baker, W.H.; Sabbe, W.E.; McCool, Y.S. Changes in Arkansas' sampled acreage testing low and high in soil phosphorus over the last 30 years. Soils and Fertility Information Article 1-93. Univ. of Arkansas Cooperative Extension Service, Little Rock, AR, 1993.
47. Sharpley, A.N.; Meisinger, J.J.; Breeuwsma, A.; Sims, T.; Daniel, T.C.; Schepers, J.S. Impacts of animal manure management on ground and surface water quality. In *Animal Waste Utilization: Effective Management of Animal Waste as a Soil Resource,* Hatfield, J., Ed. Ann Arbor Press, Chelsea, MI, 1998, pp 173-242.
48. Eck, H.; Stewart, B.A. Manure. In *Environmental Aspects of Soil Amendments,* Rechcigl, J.E., Ed. Lewis Publishers, Boca Raton, FL, 1995, pp 169-198.
49. Sharpley, A.N.; Daniel, T.C.; Sims, J.T.; Pote, D.H. Determining environmentally sound soil phosphorus levels. *J. Soil Water Conserv.* **1996**, 51, 160-166.
50. Sharpley, A.N.; Tillman, R.W.; Syers, J.K. Use of laboratory extraction data to predict losses of dissolved inorganic phosphate in surface runoff and tile drainage. *J. Environ. Qual.* **1977**, 6, 33-36.
51. Daniel, T.C.; Sharpley, A.N.; Edwards, D.R.; Wedepohl, R.; Lemunyon, J.L. Minimizing surface water eutrophication from agriculture by phosphorus management. *J. Soil Water Conserv.* **1994**, Suppl. 49, 30-38.
52. Edwards, D.R.; Daniel, T.C.; Murdoch, J.F.; Vendrell, P.F. 1993. The Moore's Creek BMP effectiveness monitoring project. Pap. 932085. ASAE, St. Joseph, MI.
53. McDowell, R.W.; Sharpley, A.N. Approximating phosphorus release from soils to surface runoff and subsurface drainage. *J. Environ. Qual.* **2001**, 30, 508-520.
54. Pote, D.H.; Daniel, T.C.; Nichols, D.J.; Sharpley, A.N.; Moore, P.A. Jr.; Miller, D.M.; Edwards, D.R. Relationship between phosphorus levels in three Ultisols and phosphorus concentrations in runoff. *J. Environ. Qual.* **1999**, 28, 170-175.
55. Torbert, H.A.; Daniel, T.C. Relating soil test P to runoff P losses in calcareous and non-calcareous soil. *In* Abstracts of Technical Papers 2001 Southern Branch of Amer. Soc. of Agron., Fort Worth TX, January 27-31, No. 28, p 9. 2001.
56. Lemunyon, J.L.; Gilbert, R.G. Concept and need for a phosphorus assessment tool. *J. Prod. Agric.* **1993**, 6, 483-486.

Chapter 17

Environmental and Agronomic Fate of Fertilizer Nitrogen

Robert G. Hoeft

Department of Crop Science, University of Illinois, 1102 South Goodwin, Urbana, IL 61801

Maintenance of a quality environment and a competitive advantage in the world market place is a goal of all that recommend or use nitrogen fertilizer. To accomplish this goal, one must fully understand the biological and chemical reactions that nitrogen undergoes in a soil system. Mineralization, the process of conversion of organic nitrogen to plant available inorganic forms is affected by climatic and prior management practices. As a result of the climatic influence, the rate of this reaction is unpredictable from year to year and thus it is difficult to predict the absolute optimum rate for any field in any year. Nitrification, the biological conversion of ammonium to nitrate is temperature dependent. Understanding this reaction allows producers to select the time of application that will minimize the potential for nitrogen to be in the nitrate form during the time period when denitrification and leaching are most likely to occur. Use of a nitrification inhibitor is another management tool that farmers can utilize to control the timing of the conversion of ammonium to nitrate. Plants recover from 30 to 40 percent of the fertilizer nitrogen in the year of application. An equal amount is converted to organic form, immobilization, and is then available for release in subsequent years through the process of mineralization. The mineralization of newly immobilized organic compounds is about 7 times faster than native organic nitrogen compounds.

Introduction

Nitrogen management is quintessential for U.S. grain producers to maintain a competitive advantage in the world market place and at the same time a quality environment. Historically, grain prices have varied considerably, depending on supply and demand. Unfortunately, in the last several years, prices have trended down, and based on world grain supply, there is little hope that they will improve in the near future. Fertilizer prices also fluctuate based on supply and demand, with supply being dictated in part by cost and availability of raw material. Over the last 20 years, U.S. farmers paid on the average 14.6 cents per pound of nitrogen for fertilizer grade anhydrous ammonia. These prices varied from a low of 11.4 to a high of 20.1 cents per pound of nitrogen. Over the same time period, corn prices received by farmers varied from $1.54 to $3.30 with an average of $2.42 per bushel. Over the last five years, the average corn price has been $2.08 and fertilizer prices 17.1 cents per pound. Even though there has been a squeeze between fertilizer and grain prices, the use of nitrogen fertilizer is still very beneficial. Assuming that all producers used the optimum rate of nitrogen, the net value from increased production associated with nitrogen fertilizer use in Illinois would have been $680 million.

At the same time as economics are becoming tighter, pressure to improve nitrogen management because of environmental concerns are being stepped up by regulatory agencies. The Mississippi River/Gulf Hypoxia Watershed Nutrient Task Force reaffirmed the commitment to reduce N loss to the Mississippi River by 30%, with some suggesting that most of the gain will come from reduction of fertilizer use. While this is an amiable goal, the relationship between fertilizer sales and the size of the hypoxia zone is not strong (Figure 1).

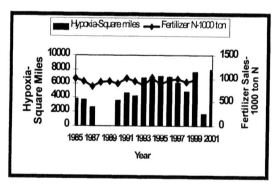

Figure 1. Size of the hypoxia zone and Illinois nitrogen fertilizer sales. Hypoxia data provided by N.N. Rabalais, R.E. Turner, and W.J. Wiseman, Jr.

To meet the goals of having an economically and environmentally sound crop production system requires the use of Best Management Practices (BMP) for nitrogen management. These practices include setting the correct rate of application, taking credit for naturally produced nitrogen, and applying fertilizers at the correct time to avoid nitrogen loss. Use of the current scientific understanding of the fate of nitrogen in soil- the nitrogen cycle (Figure 2) provides guidance for developing BMP's.

Nitrogen Cycle

Mineralization

Mineralization, the microbial process that results in the conversion of organic nitrogen to inorganic nitrogen (plant available nitrogen), varies from year to year due to climatic variation. Under warm, moist conditions, the average rule of thumb is that there will be approximately 23 kg/ha of nitrogen released for each 1 percent organic matter. However, some of our recent work has shown that this can vary by at least 2 fold from year to year and as much as 4 fold depending on past management.

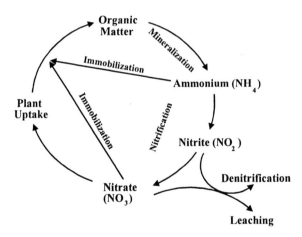

Figure 2. The Nitrogen Cycle

This differential in nitrogen release created by differential nitrogen management in the past affects the amount and relative proportion of nitrogen taken up by plants from soil and current fertilizer application (Figure 3). The data provided below comes from a study in which the nitrogen rates listed were applied each year for 15 years prior to collection of the data in the figure. During that time, the excess nitrogen from fertilizers obviously resulted in a build-up of easily mineralizable nitrogen that was released and taken up by the corn plants. In addition, the higher the nitrogen rate in the year of application, the greater the fertilizer nitrogen uptake by plants. In terms of fertilizer recovery, the greatest recovery occurred at the rate that was near the optimum for crop production, approximately 160 kg/ha of nitrogen.

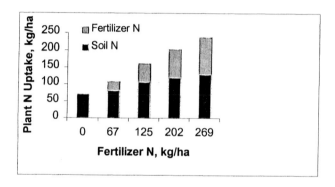

Figure 3. Relative source of corn N uptake.

Nitrification

Nitrification is the biological conversion of ammonium to nitrate with an intermediate production of nitrite. As shown in figure 2, plants can utilize both ammonium and nitrate nitrogen. However, the majority of the nitrogen taken up by plants is in the nitrate form. The rate of nitrification is temperature dependent (Figure 4) and is affected by the addition of a nitrification inhibitor (Figure 5). An understanding of the nitrification process is important in terms of management of fertilizer application to avoid the potential for nitrogen loss.

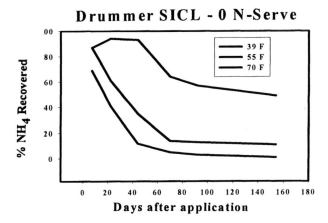

Figure 4. Disappearance of ammonium nitrogen from soils over time at different temperatures.

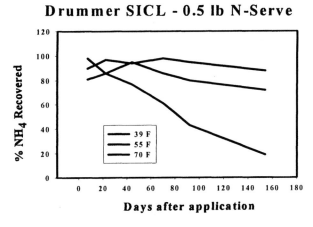

Figure 5. Disappearance of ammonium nitrogen from soils over time at different temperatures when nitrapyrin is included with the fertilizer.

Denitrification and leaching

Once it reaches the nitrate form, nitrogen is susceptible to loss by the processes of denitrification or leaching. Denitrification is most likely to occur on medium to heavy textured soils, whereas leaching is most likely to occur on lighter textured (sandy) soils. Neither of these loss mechanisms will occur unless the soils are excessively wet. In the case of denitrification, the soils must be saturated for a period of at least 3 days under warm moist conditions. Torbert et al., 1992 reported nitrogen loss values ranging from 2 to 10 percent of the fertilizer applied for each day the soils are saturated. While leaching is more of a problem on sandy soils, it does occur in heavier soils and is accentuated by the use of tile drain systems. The amount of nitrogen lost via tile line leaching is influenced by rate of nitrogen application and by the amount of water deposited on the land. In an excessively wet year (1999), nitrogen loss was equivalent to over 30 percent of the amount of fertilizer nitrogen applied, but in a dry year (2000) on the same fields, this loss was reduced to an equivalence of less than 5 percent of the fertilizer nitrogen applied (Figure 6).

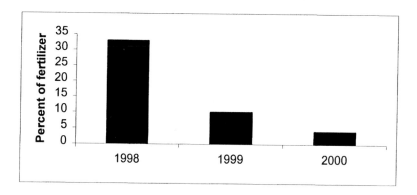

Figure 6. Percent of applied fertilizer lost from tile lines at 11 experimental locations over 3 years

Immobilization

Immobilization is the biological process of converting inorganic nitrogen to organic nitrogen. From 30 to 50 percent of the fertilizer nitrogen is immobilized (converted to organic nitrogen) during the growing season in which it was applied (Figure 7). This resulting organic nitrogen material will not be available for loss via either denitrification or leaching until it has been nitrified.

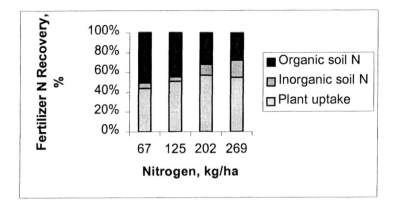

Figure 7. Fate of fertilizer nitrogen during the first growing season after application Stevens et al., 1997

Best Management Practices

Proper nitrogen rate

Since there are several biological reactions that influence the availability of nitrogen for crop use, it is difficult to establish a nitrogen rate that will be accurate for every field in every year. The economical optimum nitrogen rates varied from 80 to 260 kg/ha over the 19 years of a study at the Northwestern Illinois Research and Demonstration Center (Figure 8). This difference in response across years was in large part due to climatic differences. The years of low nitrogen need were generally characterized as being years of good mineralization or years of low yield. In contrast, the years in which the nitrogen rate required was high were generally characterized as being years of low mineralization or high nitrogen loss due to denitrification. Results similar to this occur regularly in most fields across the Corn Belt. Therefore, producers are forced to use a rate that will over the long run give them an economic optimum return. Current University of Illinois recommendation for the field used in the experiment for Figure 8 would have been approximately 170 kg/ha, a level that would have resulted in less than optimum yield in but a few of the years of the study.

Long term use of nitrogen rates in excess of those recommended will result in a marked increase in the loss of nitrogen from tile lines (Figure 9). Irrespective of the historical rate of nitrogen used, losses will be greatest in the year in which the nitrogen has been applied—the corn year.

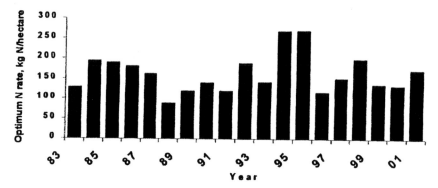

Figure 8. Variation in optimum nitrogen rate over time.

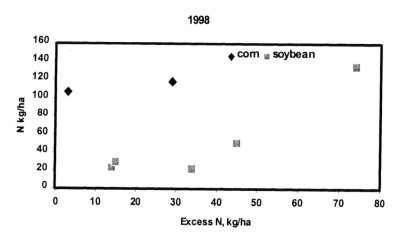

Figure 9. Relationship between historical nitrogen use as related to recommended rate and quantity of nitrogen lost from tile line.

Application time

Time of nitrogen application can have a significant impact on the potential for nitrogen loss and consequently on the recovery of fertilizer nitrogen in grain. Since nitrogen loss to the environment is limited for the most part to losses that occur when soils are warm and excessively wet, the objective of a nitrogen application plan is to apply it when the conversion of ammonium to nitrate will be slow or to use a product that will slow the conversion until after the period of excess precipitation, usually late spring to early summer. Fall applied nitrogen should be limited to sites that will freeze for a significant portion of the winter and should not be done until temperatures fall below 10 C if a nitrification inhibitor is not used or 16 C if a nitrification inhibitor is used. While the nitrification process does not stop until soil temperatures reach freezing, starting at 10-16 C is suggested as there is usually a short time period between these temperatures and freezing. Use of such management techniques have been shown to stabilize yield and minimize nitrogen loss to tile lines (Randall and Mullin, 2001).

References

1. Randall, Gyles W. and Mullin, David J. 2001. Nitrate nitrogen in surface waters as influenced by climatic conditions and agricultural practices. J. Env. Qual. 30:337-344.

2. Stevens, W.B., R.G. Hoeft, and R.L. Mulvaney. 1997. Effect of N fertilization on accumulation and release of readily-mineralizable organic N. *In* R.G. Hoeft (ed) Proc. IL. Fert. Conf. pp 65-78.

3. Torbert, H.A., R.G. Hoeft, R.M. Vanden Heuvel, and R.L. Mulvaney. 1992. Effect of moisture regime on recovery and utilization of fertilizer N applied to corn. Commun. Soil Sci. Plant Anal : 23 : 1409-1426.

Chapter 18

Working Together to Make the U.S. Environmental Protection Agency Nonpoint Source Program Effective and Efficient

Thomas E. Davenport

Water Division, U.S. Environmental Protection Agency, 77 West Jackson Boulevard, Chicago, IL 60604

The U.S. Environmental Protection Agency (EPA) and State water quality agencies are placing increasing emphasis on addressing the problems caused by nonpoint source pollution, which is now the leading cause of water pollution in the United States. The most significant nonpoint source pollutants are nutrients and sediments. The primary sources of nutrients are point sources associated with municipal wastewater discharges and nonpoint source discharges from agricultural activities. The primary agricultural sources of nutrients which can contribute to nonpoint source pollution are animal manure and commercial fertilizer. EPA, either directly or through State agencies, implements several programs under the Clean Water Act that attempt to address this issue, including the development of water quality standards for nutrients; implementation of State nonpoint source programs; point source regulations for concentrated animal feeding operations; and the development and implementation of plans to address "TMDLs"-total maximum daily loads designed to assure that water quality standards are not violated.

Introduction

Nitrogen is an essential element for life, but when present in excess within an ecosystem it most likely behaves as is a pollutant. An excess amount of nitrogen in the ecosystem is usually the result of human activity. Excessive nitrogen loading to aquatic ecosystems can result in an increase in macrophytes or phytoplankton, leading to decline of oxygen in the water column, imbalance of aquatic species, public health risks, and an overall decline of the aquatic resource (1). EPA is working with its partners to develop and implement programs to more efficiently manage nitrogen inputs and prevent environmental losses. There are several problems with addressing nutrient pollution; lack of site specific information, lack of program coordination and integration, and the lack of political will to address the issues.

Background

The lack of a comprehensive database limits our ability to accurately estimate water quality impairments in the US. However, the Clean Water Action Plan: Restoring and Protecting America's Waters (1) indicates that over-enrichment of waters by nutrients (nitrogen and phosphorus) is the biggest source of impairment of US waters. The recent State and tribal water quality data (2) indicates that approximately 40 per cent of our Nation's waters have been assessed. Since only a small portion of the Nation's waters are assessed and there are inconsistencies between states in terms of methods there is no reliable method to calculate National estimates. The lack of adequate water data quality data is a limiting factor in EPA's ability to address nutrient related issues. The percentage of waters assessed ranges from 23% for rivers and streams to 42% for lakes. These State and Tribal water quality assessments (2) indicate that:

- 35 % of the river and stream miles assessed are polluted;
- 44 % of the estuaries square miles assessed are polluted; and
- 45 % of the lake acres assessed were polluted.

Waters are considered polluted when one or more designated uses are impaired.

The leading pollutants for rivers and streams are: siltation, pathogens and nutrients. For lakes they are: nutrients, metals and siltation. The three major pollutants impacting estuaries are pathogens, organic enrichment and metals (2). The leading sources of the pollutants for rivers and lakes are the same; agriculture, hydromodification, and urban runoff (2). For estuaries the leading sources identified were municipal point sources, urban runoff and atmospheric deposition (2). The overall percentage of pollution attributable to agriculture ranges from 60 to

65% and urban runoff from 15 to 20% depending upon the reference (2, 3). Rather than looking at the percentage of assessed waters impacted by a source it is more important to look at pollution production of a source (unit area production). On a unit area basis, agriculture produces about 1% of the pollution per 1% of agricultural land use in the United States or 1:1; on the same scale, for urban land use the range is from 7.5:1 to 10:1 (3). Not only is there variability in the unit area production of pollution by land use the concentrations in runoff from the various land use varies. The range in nitrogen concentrations in urban runoff is 3 to 10 mg/L, agriculture is 0.77-5.04 mg/L, livestock operations its 6-800 mg/L (4). In addition the types of pollutants are considerably different. In urban runoff pathogens and metals the concern, for agricultural runoff sediment and nutrients are the major concerns (2).

Two examples of bodies of water impacted by nutrient over enrichment are the Dead Zone in the Gulf of Mexico, and Lake Champlain in the Northeast (4). The Dead Zone and Lake Champlain are also examples of where States and their Federal partners are actively working together to reduce nitrogen loadings as a result of a documented water quality impairment.

Dead Zone - Gulf of Mexico. Since the early 1990's a zone depleted of oxygen the size of New Jersey has been an emerging issue for the Gulf of Mexico ecosystem coastal zone near the discharge points of Mississippi/Atchafalaya Rivers. The "Dead Zone" lies along the bottom of shallow waters near the Mississippi and Atchafalaya River deltas. The cause of this dead zone has been identified as excessive nitrogen and a number of contributing factors such as the loss of wetlands (5). The Gulf fishery in the area has been devastated (4). Annually, the Mississippi River discharges approximately 900 tons of nitrate, 110,000 tons of phosphorus and 231 million tons of sediment (4). The primary source of these pollutants has been identified as nonpoint source or diffuse pollution from agricultural and urban lands (4). Thirty one States and their respective Federal and Tribal Partners are working on developing recommendations for basin wide solutions to the problem of excessive nitrogen loading. These basin wide solutions will then be translated into local actions at the State and sub watershed level.

Lake Champlain - New York and Vermont. Excessive loadings of phosphorus are causing algae blooms in parts of Lake Champlain (4). These algae blooms are impairing recreation, and reducing oxygen levels in a number of areas so as not to support fish. Algae blooms can contribute to fish kills, drinking water, livestock poisoning and increased health risk to people. New York and Vermont are working with the United States Department of Agriculture (USDA) and USEPA to implement basin strategies to reduce excessive nutrient loadings from agricultural lands.

While there are several causes of nutrient loss from agriculture, the major cause of problems like the Lake Champlain and Gulf Mexico is excessive loss of nutrients from cropland due to the over application of fertilizer (4,5). Mike Hirschi, University of Illinois agricultural engineer, (personal communication) linked the over application of fertilizer in two Illinois watersheds to elevated stream loading estimates. Hirschi estimated farmers in the Big Ditch Watershed (Illinois) annually applied 53 lb/ac per year in excess of nitrogen corn crop requirements, over a three year period this resulted in almost 3 million pounds of excess nitrogen being over applied to crops. Correspondingly almost 3 million pounds of nitrogen was measured at the watershed's outlet during the same time frame (6). In the Vermillion River Watershed (Illinois) a similar trend was documented, over fertilization of cropland and elevated stream loadings (6). These are just a few examples of prevalent nonpoint source nutrient related problems facing the Nation today. While extensive data is lacking, States and Tribes report there are many localized nonpoint source nutrient related pollution problems needing to be addressed in order to restore water quality.

Programs

There are a number of federal tools available to address nonpoint source issues. The principal programs are the Clean Water Act and Farm Bill. The Farm Bill is under the jurisdiction of the USDA and the Clean Water Act is implemented by the USEPA and its partners. The majority of USEPA's Programs are implemented through State and Tribal governments. There are number of programs available within the Clean Water Act to help address nutrient problems. The main programs applicable across the Nation are: Water Quality Standards (section 303(c)), Water Quality Management (Section 303), Clean Lakes (Section 314), Nonpoint Source Management (Section 319) and National Pollutant Discharge Elimination System (NPDES) (section 402). A number of these programs complement other federal and state programs aimed at managing nutrients. While the programs receiving the most attention are the Total Maximum Daily Load (TMDL) (section 303(d)) and nonpoint source (section 319), they are just components within the overall framework to address water quality issues. Figure 1 presents the water quality management framework for protecting and restoring water quality. The first step in the process is the establishment of water quality standards.

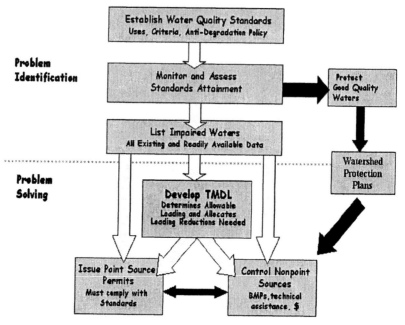

Figure 1: Water Quality management framework.

Water Quality Standards

Water quality standards are the foundation for water quality management in the United States. Water quality standards consist of four components: designated use, numeric criteria, narrative criteria and an anti-degradation clause. Achieving and maintaining compliance with water quality standards are driving forces for the Clean Water Act programs. Water quality standards play a critical role in defining problem areas and setting pollution reduction goals. The lack of adequate or subjective nutrient standards hinders water pollution agencies ability to identify and address nutrient related problems.

To protect and manage the Nation's waters, USEPA is working with states and designated tribes to develop and adopt numeric criteria for nutrients that are tailored to reflect the different types of water bodies and the different ecoregions around the country. The goal is to establish an objective, scientifically sound basis for assessing nutrient over enrichment problems. The USEPA has allowed states and tribes flexibility to modify or improve on the basic approach of utilizing nutrient ecoregions based natural and anthropogenic factors to set their standards. For example in the Upper Midwest a refinement of the recommended approach has been adopted. In this Upper Midwest approach the relative importance of various environmental characteristics affecting nutrient concentrations by the use of regression tree analysis will be incorporated into the water quality standards setting process. The areas or regions, within a state or reservation, will then be defined based only on the most statistically significant characteristics.

TMDL

TMDLs are not limited to just waters impaired by nutrient over enrichment; they are targeted to water bodies not meeting their water quality standards or projected not to meet standards with technology based controls in place. States, territories, and authorized tribes are to establish Total Maximum Daily Loads (TMDL) that will meet water quality standards for each of its listed waters. A TMDL is the sum of the individual wasteload allocations for point sources and load allocations for nonpoint sources (NPSs) and natural background (40 CFR 130.2) with a margin of safety (CWA Section 303(d)(1)(C)). The TMDL can be generically described by the following equation:

$$TMDL = LC = \Box WLA + \Box LA + MOS$$

where: LC = loading capacity that a water body can receive without violating water quality standards;
WLA= wasteload allocations for individual point sources;
LA= load allocation of NPSs;
MOS= margin of safety and in most cases background

Once listed by an entity the water bodies are prioritize for TMDL development based upon a number of factors. The water body remains on the 303(d) list until it meets water quality standards. An implementation plan needs to be developed to address the needed nonpoint source loading reductions under the LA. The implementation plan needs to be developed utilizing the program neutral planning process currently being developed jointly by USDA and USEPA. Presently, the only enforceable component of the TMDL is the WLA which is implemented under the NPDES program.

Clean Lakes

Section 314 is probably the most effective National watershed/water quality program the USEPA has ever been involved in (7). Structurally, the program is simple: identify priority lakes for individual project action, select a lake project, complete a diagnostic/feasibility study on the lake and its watershed, implement the results of the study and then evaluate the impact of what was implemented. The priority list of lakes serves as basis to identify possible lake projects; most states utilized criteria to select projects that included public benefits and willingness to proceed factors. The priority process in most states focused on; lakes that were impaired, maximum public benefit, lake owners and watershed residents commitment, and technically feasible to correct the impairment. For a selected lake project, a diagnostic/feasibility study that would document problems, causes and sources of in-lake impairments and develop a number of alternatives to address the problems would be completed. The public would be involved in the study and assist not only in identifying and selecting management alternatives, but also in the problem identification phase. The selected alternative is then implemented; when all watershed related activities have been implemented the in-lake management techniques would be scheduled. Then a few completed projects would be selected for a follow up intensive evaluation monitoring efforts 3-5 years after the watershed treatment/in-lake techniques have been completed. There are a number of successful projects such as Lake LeAquaNa and Johnson Sauk Lake where a combination of funding sources supported the implementation of an approved diagnostic/feasibility study to address nutrient enrichment problems and restore water quality (9). Unfortunately, at the present time USEPA receives no funding from Congress to support this balanced watershed/lakes program, However, USEPA encourages States to continue to implement the Clean Lakes Program through their State Nonpoint Source Management Program in order to support lake watershed based projects in efficient manner. In addition, this multi phase process closely follows the TMDL process mentioned earlier.

Nonpoint Source Program

Section 319 establishes a National framework for States (and approved Tribes) to create nonpoint source management programs with USEPA assistance. The section 319 program requires that States develop and gain USEPA approval of a Nonpoint Source Assessment Report that documents the nonpoint source pollution impairments and threats to its navigable waters. To be eligible for funding assistance under section 319 the States then have to develop a management program to address the pollution identified in the approved assessment report. This management framework relies heavily on two components: public involvement and watershed level management. The section 319 Program should be a key implementation tool in the TMDL process to restore water quality. The problem with this concept is, most state Section 319 Programs are not water quality based, i.e. they are not based upon attaining or maintaining water quality standards (8). Most states and eligible tribes developed voluntary watershed-based programs that focus on providing technical and financial assistance to polluters in a broad-based stewardship manner rather than water quality-based approach. The stewardship approach distributes technical and financial resources equally, rather than in a prioritized manner. The stewardship approach is relatively ineffective because the technical and financial resources are not targeted to the main pollutants of concern and their primary sources (8, 9, and 10).

In contrast, the water quality-based approach requires identification of the pollutant(s) causing the water quality impairment and then directs the focus of land management improvements to critical pollutant source areas. In this manner, the pollutant(s) source and transport to the impaired (or threatened) water resource is effectively addressed (8). A high level of treatment in critical areas will result in a greater reduction of pollutant delivery as compared to a broad brush approach with lower levels of land treatment over larger areas (8, 9, 10, and 11).

One of the main reasons for states' preference for the stewardship approach is due to it's consistency with the existing Federal and state conservation programs. Another major reason was the lack of information and methodologies to support the water quality-based approach. To overcome this lack of information and to provide a basis to guide future program direction, USEPA, in partnership with the States, Tribes and interested Federal agencies, established the Section 319 Nonpoint Source National Monitoring Program (NMP). In general, States have failed to adequately evaluate their watershed efforts due to the lack of monitoring and record keeping (8, 9, 10, and 11). Additionally, states have not supported comprehensive outreach programs to build social capacity to support behavior changes at the watershed level. Nonpoint source pollution is really a people pollution issue and people changing their behavior are a key to its long term reduction (9, 10, and 12).

National Pollutant Discharge Elimination System

Under the Clean Water Act, all point source discharges of pollution require a permit. Point source is defined as any discernible, confined, and discrete conveyance of pollutants to a water body. For all discharges, technology based controls are required except in circumstances where more stringent effluent limitations are required to meet water quality standards. In addition to industries and municipal wastewater treatment plants, NPDES also covers both urban storm water runoff and animal feeding operations that meet certain size criteria or are causing a water quality standards violation (Clean Water Act section 402). It is expected under both of these efforts there will be nutrient management requirements. The minimum storm water elements will require information and education requirements to educate land owners on the proper fertilization and management of lawns and open spaces, including parks. The enforceable mechanism for implementing water pollution control requirements is the (NPDES) permit required under the Clean Water Act (section 402).

Discussion

The development of nutrient criteria will allow us to enhance the identification and quantification of water negatively impacted by excess nutrient. Figure 1 highlights the water quality management process for addressing identified problems. In order to increase the effectiveness of the nonpoint source efforts we must improve the problem identification phase, upgrade implementation of state nonpoint source projects, and increase coordination of federal programs.

The development and adoption of nutrient water quality standards by State Water Quality Agencies is necessary to improve the problem identification of impaired waters. Even with water quality standards in place, the lack of adequate monitoring programs (as indicated by the percentage of waters assessed [2]) at the state and federal level will continually limit the states ability to identify problem areas. The recent increases in the Clean Water Act operating funds for states should be targeted to water quality monitoring in order to increase the percentage of waters assessed. Enhancement of state monitoring efforts should focus on three areas: state wide (for identifying problems), watershed (to document the impact of management), and best management practice levels.

In my opinion, with improved water quality monitoring approaches states could ensure restoration and/or protection projects are targeted to priority areas and begin to make differences in protecting and restoring water quality. With improved problem identification, states can then move their Section 319 Programs from stewardship-based to water quality-based efforts. Once states start funding water quality based projects, they will have to change how section 319 is implemented at the watershed level. The NMP has provided a wealth of information regarding project level enhancements for section 319 projects. The following are recommendations from NMP Successes and Recommendations Document (8).
Project Organization And Administration:

- ☐ Clearly define roles and responsibilities of federal, state/regional, and local governments for effective interagency coordination and cooperation.
- ☐ Involve all major agencies and landowners in BMP selection and planning to maintain long-term commitments.
- ☐ Close coordination is needed between monitoring/evaluation and land treatment implementation agencies and personnel.
- ☐ Ensure up-front commitment of funds for the multi-year project period. Due to the long-term nature of watershed projects, reliable funding is needed to facilitate long-term planning and budgeting. A short-funding cycle or a requirement to request and compete for funds annually does not ensure full comprehensive implementation of project activities, continuity of project staff, and reduces the effectiveness of projects. The short funding cycle is particularly damaging to monitoring and outreach efforts.

Land Treatment Implementation
- Implement appropriate and sufficient BMPs that address the water quality problem. A high level of BMP implementation is needed because it is necessary to affect changes of at least 20 percent in the water quality pollutant levels or loads before statistical linkage can be made. Implementation should focus on critical areas.
- Target BMP implementation to the critical pollutant source areas and pollutants, to reduce the delivery of the pollutants to the water quality resource of concern.
- Provide long-term operation and maintenance (O&M) of BMPs for both management and structural BMPs. Questions of who is responsible for O&M need to be addressed up front for all parties involved in the project and be documented in the implementation plan.
- Employ systems of BMPs. The installation of one structural or management BMP is rarely sufficient to entirely control the pollutant of concern. Combinations of BMPs that control the same pollutant are generally most effective.

Information and Education
- Provide information and education (I&E) for a high level of land-owner participation, prior to project implementation and continued throughout the project. I&E efforts conducted early in the project may be necessary to increase general awareness of the water quality problem, gain public support for the project, and improve land owner understanding of their contributions to the problem. Continuing I&E efforts are needed to assist land owners in the management and maintenance of implemented BMPs and to inform them of the impact of their actions."

The last major area needing focus is the coordination of Federal programs that can be utilized to address nonpoint source pollution. The lack of a unified strategy at the state level leads to the various Federal programs competing for the same client rather than each targeting specific audiences to maximize potential involvement. In my opinion with the development of nutrient criteria, State and Federal agencies will be able to target specific water bodies for implementation and coordinate which water bodies are addressed. The program neutral planning process will highlight needs and problems that have to be addressed to attain and maintain water quality standards. Once issues and problems are identified various State and Federal Programs can be targeted to correct them. This program specific targeting avoids duplication of effort and increase efficiency of program delivery.

References

1. Clean Water Action Plan Team. 1998. Clean Water Action Plan: Restoring and Protecting America's Waters. United States Environmental Protection Agency, Office of Water, Washington, DC 89pp.
2. United State Environmental Protection Agency (USEPA). 2000. National Water Quality Inventory: 1998 Report to Congress. EPA-841-R-00-001. USEPA, Office of Water, Washington, DC.
3. USEPA.1992. Environmental impacts of storm water discharges, a national profile. EPA 841-R-92-001. June 1992. USEPA, Office of Water, Washington, DC.44pp.
4. Coast Alliance. 2000. Pointless Pollution: Preventing polluted runoff and protecting America's coasts. Coast Alliance, Washington, DC. 42pp.
5. A Watershed Decade. EPA 840-R-00-002. 2001. USEPA, Office of Water, Washington, DC.
6. Agri-View. 1999. Nitrate in water is a matter of rates and timing. Agri-View October 28, 1999, Madison, WI
7. A Commitment to Watershed Protection: A Review of the Clean Lakes Program. EPA 841-R-93-001. 1993. USEPA, Office of Water, Washington, DC
8. Lombardo, L.A., G.L. Grabow, J. Spooner, D.E. Line, D.L. Osmond, and G.D. Jennings. 2000. Section 319 Nonpoint Source National Monitoring Program Successes and Recommendations. NCSU Water Quality Group, Biological and Agricultural Engineering Department, NC State University, Raleigh, North Carolina. 31 p.
9. Davenport, T. 2002. The Watershed Project Management Guide (in press). CRC Press, Boca Raton, FL
10. Dr. Peter Nowak, Interaction of social and bio-physical parameters: what needs to be done? Presentation at ASIWPCA/EPA Nonpoint Source Meeting, November 28, New Orleans, 2001
11. USEPA. 1993. Evaluation of Experimental Rural Clean Water Program. EPA-841-R-93-005. USEPA, Office of Water, Washington, DC
12. Davenport, T.E., 2000. Project Evaluation. Water Resource IMPACT 2(1):4-7.

Indexes

Author Index

Averitt, David W., 32, 61
Bashour, Isam, 90
Bi, S. P., 100
Boos, Dennis, 75
Chassaniol, Kirk, 3
Cisar, J. L., 161
Davenport, Thomas E., 244
Erickson, J. E., 161
Gan, N., 100
Hall, William L., Jr., 32, 61, 180
Hannoush, Ghada, 90
Haydu, J. J., 161
Hoeft, Robert G., 235
Hopwood, E. W., 180
Jackson, Peter E., 3
Jones-Lee, Anne, 207
Kane, Peter F., 61
Kawar, Nasri, 90
Labno, K. A., 112
Lauterbach, A., 45
Lee, G. Fred, 207

Littell, R. C., 180
Pan, W. L., 112
Potter, K. N., 220
Proctor, Charles, 75
Reetz, Harold F., Jr, 196
Robarge, Wayne P., 32, 75
Sartain, J. B., 180
Snyder, G. H., 161
Stevens, R. G., 112
Stewart, W. M., 151
Tang, W., 100
Thomas, Dave, 3
Torbert, H. A., 220
Urbansky, Edward Todd, 16
Volin, J. C., 161
Wang, C. Y., 100
Weber, Linda D., 32
Wen, L. X., 100
Woltering, Daniel M., 124
Xu, R., 100

Subject Index

A

Absorption, perchlorate, 29
Accelerated lab extraction
　correlation of soil incubation nitrogen release and, 191–192
　future work, 193–194
　regression analysis, 192
　See also Soil incubation methodology
Accumulation
　perchlorate, 29
　See also Cadmium accumulation
Acreage, planted, fertilizer consumption, 152, 154f
Agricultural grade production, sodium and potassium nitrates, 49–50
Agriculture. See Nutrient losses; Site-specific agriculture
Alternative landscapes, Florida Yards and Neighborhoods (FYN) program, 163–164
Aluminum speciation
　application of model to soil water samples, 107–108
　calculated, fro soil solutions of known composition, 108t
　comparing predicted vs. measured concentrations, 109t
　critical issue, 101
　distribution as function of pH with different sulfate concentrations, 105f, 106f
　distribution for soil waters in contact with jubanite, 104
　effect of mineral solubility on distribution of, 107
　effect of SO_4^{2-} concentration on distribution of, and dissolved total Al concentration, 105
　influence of pH, 104
　mass balance in equilibrium with presence of jubanite, 103t
　model theory, 102
Ammonium nitrate (AN)
　fertilizer source materials, 77t
　means of trace metals, Lebanon, 93f
　N-fertilizer in Lebanon, 91–92
　sampling schedule, 82t
　trace metal content, 83, 85
　See also Nitrate salts; Trace metal content of fertilizers
Ammonium perchlorate, perchlorate concerns, 8
Ammonium sulfate (AS)
　fertilizer source materials, 77t
　intercomparison analyses for, 88t
　means of trace metals, Lebanon, 93f
　N-fertilizer in Lebanon, 91–92
　sampling schedule, 82t
　trace metal content, 83, 84t
　See also Trace metal content of fertilizers
Analysis. See Slow-release fertilizers
Anion exchange resin, perchlorate removal, 55, 56f
Anion MicroMembrane Suppressor (AMMS), conductivity, 6
Anion Self-Regenerating Suppressor (ASRS), inorganic anions, 6, 7f, 8t
Application rate, parameter for risk based concentration (RBC), 134t
Application time, nitrogen fertilizer, 243
Arsenic

concentrations in phosphate and micronutrient fertilizers, 145t
digestion techniques, 65t
instrument techniques, 67t
metals in assessment, 128
RBC (risk based concentration) for all scenarios, 140t, 141t
RBCs for, in fertilizer, 142f
statistical summary, 68, 69t
unit RBC value in fertilizer, 143t
See also Heavy metals in fertilizers
Association of American Plant Food Control Officials (AAPFCO)
Controlled Release Fertilizer Task Force, 182
enhanced efficiency categories, 181
See also Slow-release fertilizers
Association of Analytical Chemists International (AOACI), inconsistencies in analytical methods, 181–182
Availability, nutrient in fertilizers, 17–18

B

Best management practices (BMPs)
nitrogen fertilizer, 241–243
nutrient runoff control, 215–216
Biological exposure parameters, summary intake factors (SIFs), 133t

C

Cadmium
accumulation over time, 113
concentrations in phosphate and micronutrient fertilizers, 145t
contaminant in agricultural soil, 113
digestion techniques, 65t
instrument techniques, 67t
means of trace metals in fertilizers, Lebanon, 98f
means of trace metals in N-fertilizers, Lebanon, 93f
means of trace metals in NPK-fertilizers, Lebanon, 96f
means of trace metals in P-fertilizers, Lebanon, 94f
metals in assessment, 128
ranges and means in fertilizers, Lebanon, 98t
risk based concentration (RBC) for all scenarios, 140t, 141t
statistical summary, 68, 69t
trace concentrations in K-fertilizers, Lebanon, 95t
trace concentrations in N-fertilizers, Lebanon, 92t
trace concentrations in NPK-fertilizers, Lebanon, 95t
trace concentrations in P-fertilizers, Lebanon, 94t
unit RBC value in fertilizer, 143t
See also Heavy metals in fertilizers; Trace metal content of fertilizers, Lebanon
Cadmium accumulation
fertilizer sources, annual application rates of P and Zn, Cd concentrations and Cd rates, 115t
fresh weight grain and tuber yields over one year fertilizer application, 119t
grain Cd concentrations over second year application, 118f
Idaho rock phosphate (RP) and grain Cd, 118–119
methods, 114–116
potato, 119–121
potato samples, 116
range of grain Cd concentrations, 117–118
relationship between potato tuber Cd and total soil Cd, 120f
relationship between wheat grain Cd and total soil Cd, 117f
soil sampling method, 115–116

two-year randomized complete block experiment, 114–115
wheat, 116–119
wheat sample, 116
Caliche ore
analysis of nitrate ore samples, 48t
background and brief history, 46
location in northern Chile, 47f
origin of deposits, 48–49
SQM, 46
use of SQM nitrates in United States, 47–48
See also Nitrate fertilizers; Sociedad Química y Minera S.A. (SQM)
California Department of Food and Agriculture (CDFA)
risk assessment, 125–126
See also Health risk assessment
Chemical risk assessment
fertilizers, 125
See also Health risk assessment
Chilean nitrate, perchlorate, 33
Chilean nitrate products, crops, 30
Chile saltpeter
nitrogen sources, 19–20
perchlorate analysis in, 12
Chromium
concentrations in phosphate and micronutrient fertilizers, 145t
means of trace metals in fertilizers, Lebanon, 98f
means of trace metals in N-fertilizers, Lebanon, 93f
means of trace metals in NPK-fertilizers, Lebanon, 96f
means of trace metals in P-fertilizers, Lebanon, 94f
metals in assessment, 128
ranges and means in fertilizers, Lebanon, 98t
risk based concentrations (RBC) for all scenarios, 140t, 141t
trace concentrations in K-fertilizers, Lebanon, 95t
trace concentrations in N-fertilizers, Lebanon, 92t
trace concentrations in NPK-fertilizers, Lebanon, 95t
trace concentrations in P-fertilizers, Lebanon, 94t
See also Trace metal content of fertilizers, Lebanon
Citrus fruits, Chilean nitrate products, 30
Clean Water Act
National Pollutant Discharge Elimination System (NPDES), 251
water quality, 252
Cobalt
concentrations in phosphate and micronutrient fertilizers, 145t
means of trace metals in fertilizers, Lebanon, 98f
means of trace metals in N-fertilizers, Lebanon, 93f
means of trace metals in NPK-fertilizers, Lebanon, 96f
means of trace metals in P-fertilizers, Lebanon, 94f
metals in assessment, 128
ranges and means in fertilizers, Lebanon, 98t
risk based concentrations (RBC) for all scenarios, 140t, 141t
trace concentrations in K-fertilizers, Lebanon, 95t
trace concentrations in N-fertilizers, Lebanon, 92t
trace concentrations in NPK-fertilizers, Lebanon, 95t
trace concentrations in P-fertilizers, Lebanon, 94t
unit RBC value in fertilizer, 143t
See also Trace metal content of fertilizers, Lebanon
Columns, ion chromatography
advances, 4
Compliance monitoring, drinking water standards, 6
Concentrated animal feeding operations (CAFOs)
manure, 230

nutrient management, 197
Conductivity, detection for ion chromatography, 5, 6
Conservation tillage
  effect of tillage system on PO_4-P content of runoff, 224f
  influence on runoff nutrient losses, 223t
  reducing non-point source pollution from crop land, 221–222
  soil moisture and tillage system, 225–226, 230
  See also Nutrient losses
Consumption, American fertilizer, 20
Contaminant, perchlorate, 8–9
Controlled Release Fertilizer Task Force
  activities, 182
  See also Slow-release fertilizers
Copper
  concentrations in phosphate and micronutrient fertilizers, 145t
  metals in assessment, 128
  risk based concentration (RBC) for all scenarios, 140t, 141t
Corn yields, influence of soil test K levels, 203, 204f
Crop production, site-specific systems, 197
Crystallization temperature, sodium nitrate, 50

D

Dead Zone, Gulf of Mexico, nutrient over enrichment, 246
Denitrification, nitrogen fertilizer, 240
Dermal absorption factor, calculating summary intake factors (SIFs), 133t
Detection, ion chromatography advances, 4
Development. See Slow-release fertilizers
Di-ammonium phosphate (DAP) fertilizer source materials, 77t
  intercomparison analyses, 87t
  sampling schedule, 82t
  trace metal content, 82, 83t
  See also Trace metal content of fertilizers
Drinking Water Standards, inorganic anion analysis, 6

E

Economics
  nitrogen fertilizer, 236
  P and K fertilization, 205
  site-specific agriculture, 201
Education, nonpoint source programs, 253
Efficiency, site-specific systems, 202, 205
Environmental impact, site-specific systems, 202, 205
Environmental Protection Agency (EPA). See Nonpoint source pollution; United States Environmental Protection Agency (EPA)
Eutrophication
  waterbodies, 208
  See also Excessive fertilization
Excessive fertilization
  allowable nutrient load to waterbodies, 217
  approaches to developing nutrient criteria, 212f
  control of phosphorus and nitrogen, 215–216
  nitrogen to aquatic ecosystems, 245
  nutrient criteria, 211, 213
  nutrients of concern, 209–210
  phosphorus index, 210–211
  recommendations, 217–218
  total P vs. algal-available phosphorus, 210
  waterbodies, 208
  water quality impacts of waterbody, 208–209

See also Nutrient management; Phosphorus

F

Farming, high-tech, nutrient runoff management, 216
Fate, nitrogen fertilizer, 241*f*
Fertilization
 effects on nitrogen, 172–173
 excessive, 208
 See also Excessive fertilization; Nitrogen leaching and runoff
Fertilizer Regulation Act, passage, 114
Fertilizers
 agricultural, and perchlorate, 18
 American consumption, 20
 analysis of real world samples, 27–28
 annual consumption/application of nitrogen, for several states, 21*t*
 application and nutrient loss, 222, 225
 application of perchlorate-containing, 30
 commercial lawn and garden, and perchlorate, 41, 42*t*
 complexation electrospray ionization mass spectrometry (cESI–MS), 25
 concentrations in components, 34*t*
 consumption of nitrate salts in regions of continental United States, 21*t*
 corrected perchlorate concentrations in components, 36*t*
 evaluation of participant laboratories, 26–27
 implications for vascular plants, 28–30
 initial investigations of, for perchlorate occurrence, 23–25
 interlaboratory corroboration, 24–25
 ion chromatography (IC), 25
 issues for trace analysis, 24
 lifecycle, 127*f*
 manure, 230
 mechanisms to delay nutrient release, 17
 nitrate salts, 19
 nitrogen sources, 19–20, 22
 nutrient availability, 17–18
 perchlorate concentrations in commercial, 35*t*
 perchlorate in environment, 16–17
 phosphate sources, 22
 potassium sources, 22–23
 production recordkeeping, 23
 RBCs (risk based concentrations) for arsenic in, 142*f*
 RBCs in setting standards, 143–144
 sampling, 25
 screening survey, 113
 sodium and potassium nitrates, 20, 22
 sodium nitrate, 19–20
 source materials by IMC Global, 41
 survey, 26–30
 sylvinite deposits, 23
 unit RBC values for metals in, 143*t*
 See also Health risk assessment; Heavy metals in fertilizers; Nitrate fertilizers; Nitrogen fertilizers; Nutrient losses; Perchlorate; Slow-release fertilizers; Trace metal content of fertilizers; Trace metal content of fertilizers, Lebanon
Florida Yards and Neighborhoods (FYN) program
 alternative landscapes, 163–164
 construction of experimental facility, 164
 facility assessing runoff and leaching, 166*f*
 See also Nitrogen leaching and runoff
Foundation of Agronomic Research (FAR), 197
Fraction of land, parameter for risk based concentration, 135*t*

G

Grain. *See* Cadmium accumulation
Grid sampling, site-specific systems, 198–199
Grid size, site-specific systems, 199
Gulf of Mexico Dead Zone, nutrient over enrichment, 246

H

Health concern, perchlorate, 8
Health risk assessment
  back-calculated risk based approach by The Fertilizer Institute (TFI) and California Department of Food and Agriculture (CDFA), 130
  conceptual model, 128–130
  Development of Risk-Based Concentrations for Arsenic, Cadmium and Lead in Inorganic Commercial Fertilizers (CDFA), 126
  Estimating Risk from Contaminants Contained in Agricultural Fertilizers (U.S.EPA), 125–126
  Health Risk Evaluation of Select Metals in Inorganic Fertilizers Post Application (TFI), 126
  human health risk equation, 130
  initial screening by U.S.EPA and CDFA, 128
  inverse relationship correlation between soil:water partitioning coefficients (K_d) and plant uptake factors (PUFs), 138f
  lifecycle of fertilizer, 127f
  metal concentration in phosphate and micronutrient fertilizer products, 145t
  metals for evaluation in TFI, 128
  methodology, 130–138
  most sensitive parameters in model, 136–138
  numerical values for risk based concentration (RBC) equation, 132
  parameters to calculate RBCs, 134t, 135t
  parameters to calculate summary intact factors (SIFs), 133t
  plant uptake factor (PUF), 136, 137–138
  products with potential for soil loading of metals, 127–128
  RBC, 129–130
  RBC equation, 131–132
  RBC equation for multi-crop farm scenario, 132
  RBC for arsenic in fertilizer, 142f
  RBCs for all scenarios (lower bound conditions), 141t
  RBCs for all scenarios (upper bound conditions), 140t
  RBCs for twelve metals in fertilizers, 138–139
  soil:water partitioning coefficients (K_d), 136–137
  unit RBC calculation, 132, 136
  unit RBC values for metals in fertilizers, 143t
  use of RBCs in evaluating product safety and setting standards, 143–144
Heavy metals in fertilizers
  conflicting information, 64
  digestion techniques, 65t, 66t
  EPA 3050B, 63
  EPA 3051A, 63
  EPA 3052, 63
  EPA methods, 63
  future work, 72
  instrument techniques, 67t
  public concern, 62
  Statement of Uniform Interpretation and Policy 25 (SUIP25), 62
  statistical summary for cadmium, lead, and arsenic, 68, 69t
  statistical summary for selenium and mercury, 70t, 71
  survey design, 64

survey of state regulatory agencies, 62–63
survey participants, 73–74
High performance liquid chromatography (HPLC), ion chromatography (IC), 5
High-tech farming, nutrient runoff management, 216

I

IMC Global
  addressing perchlorate contamination, 42
  fertilizer source materials by, 41
  perchlorate by ion chromatography, 39
  See also Perchlorate
Immobilization, nitrogen fertilizer, 240, 241f
Incubation. *See* Soil incubation methodology
Index of crop prices, fertilizer consumption, 152, 156f, 157
Index of fertilizer prices, fertilizer consumption, 152, 155f, 157
Information, nonpoint source programs, 253
Inhibitor, nitrification, 238, 239f
Inhibitor materials, enhanced efficiency, 181
Inorganic nutrient use
  consumption of nitrogen, phosphate, and potash, 152, 153f
  effect of planted crop acres, 152, 154f
  estimated total nutrient removal relative to, 158f
  index of crop price paid to farmers, 152, 156f, 157
  index of fertilizer price paid by farmers, 152, 155f, 157
  influencing factors, 152
  nutrient removal/use ratio, 157, 159

Instrumentation, ion chromatography advances, 4, 5
Ion chromatography (IC)
  advances, 4
  analysis of ionic species, 3
  analysis of perchlorate in high ionic strength matrices, 12
  analytical capability for perchlorate, 39
  analytical method, 38
  analytical protocols, 37–39
  anionic exchange column with Anion Self-Regenerating Suppressor (ASRS), 6, 7f
  calibration curves, 38f
  common inorganic ion analysis, 6
  comparison of AS-11 and AS-16 columns, 37f
  compliance monitoring, 6
  conductivity as detection method, 5, 6
  determination of perchlorate in Chile saltpeter extract, 12f
  determination of perchlorate in reclaimed wastewater, 13, 14f
  determination of perchlorate using AS16 column, 10f
  determination of trace perchlorate, 9
  drinking water standards, 6
  equipment, 37
  ground water samples, 9, 11
  Method 300.0 with AS14A column and ASRS, 8t
  Method 314.0 for perchlorate analysis, 9, 13, 15
  method detection limit and calibration, 38–39
  method detection limits, 5
  perchlorate analysis, 8–13, 33
  perchlorate in presence of sulfate, 11f
  principles, 5
  regulatory methods, 4t
Ion exchange, perchlorate removal, 55, 56f
Irrigation, alternative plant materials, 163

Irrigation water, source of perchlorate, 30

J

Johnson Sauk Lake, nutrient enrichment problems, 250
Jubanite [Al(SO$_4$)(OH)·5H$_2$O]
  distribution of Al speciation for soil waters in contact with, 104
  formation and presence, 101
  mass balance in equilibrium with presence of solid phase, 103t
  model theory, 102
  See also Aluminum speciation

L

Laboratories, participant, perchlorate detection, 26–27
Lake Champlain, nutrient over enrichment, 246–247
Lake LeAquaNa, nutrient enrichment problems, 250
Landscapes
  alternative plant materials, 163
  Florida Yards and Neighborhoods (FYN) program, 163–164
Land treatment, nonpoint source programs, 253
Langbeinite ore, fertilizer source materials, 41
Leaching
  nitrogen fertilizer, 240
  See also Nitrogen leaching and runoff
Lead
  concentrations in phosphate and micronutrient fertilizers, 145t
  digestion techniques, 65t
  instrument techniques, 67t
  means of trace metals in fertilizers, Lebanon, 98f
  means of trace metals in N-fertilizers, Lebanon, 93f
  means of trace metals in NPK-fertilizers, Lebanon, 96f
  means of trace metals in P-fertilizers, Lebanon, 94f
  metals in assessment, 128
  ranges and means in fertilizers, Lebanon, 98t
  risk based concentrations (RBC) for all scenarios, 140t, 141t
  statistical summary, 68, 69t
  trace concentrations in K-fertilizers, Lebanon, 95t
  trace concentrations in N-fertilizers, Lebanon, 92t
  trace concentrations in NPK-fertilizers, Lebanon, 95t
  trace concentrations in P-fertilizers, Lebanon, 94t
  unit RBC value in fertilizer, 143t
  See also Heavy metals in fertilizers; Trace metal content of fertilizers, Lebanon
Lebanon
  average annual fertilizer application rates, 96
  fertilizers in Lebanese market, 90
  public concern for trace metals in fertilizers, 90
  See also Trace metal content of fertilizers, Lebanon
Lifecycle, fertilizer, 127f

M

Magruder check samples, perchlorate content, 39, 40t
Manure, nutrient losses, 230
Mechanisms
  acidic deposition, 101
  delaying nutrient release, 17
Mercury
  concentrations in phosphate and micronutrient fertilizers, 145t

digestion techniques, 66*t*
instrument techniques, 67*t*
metals in assessment, 128
risk based concentration (RBC) for all scenarios, 140*t*, 141*t*
statistical summary, 70*t*, 71
unit RBC value in fertilizer, 143*t*
See also Heavy metals in fertilizers
Metal content. *See* Heavy metals in fertilizer; Trace metal content of fertilizers; Trace metal content of fertilizers, Lebanon
Metals. *See* Health risk assessment
Methods
regulatory for ion chromatography, 4*t*
See also United States Environmental Protection Agency (EPA)
Micronutrient fertilizer, metal concentrations, 145*t*
Mineralization, nitrogen fertilizer, 237–238
Mineral solubility, effect on distribution of Al speciation, 107
Model
nutrient load to waterbodies, 217
risk assessment, 128–130
soil water samples and Al speciation, 107–108
See also Aluminum speciation; Risk based concentrations (RBCs)
Molybdenum
concentrations in phosphate and micronutrient fertilizers, 145*t*
metals in assessment, 128
risk based concentration (RBC) for all scenarios, 140*t*, 141*t*
unit RBC value in fertilizer, 143*t*
Mono-ammonium phosphate (MAP)
fertilizer source materials, 77*t*
intercomparison analyses, 87*t*
sampling schedule, 82*t*
trace metal content, 82, 83*t*
See also Trace metal content of fertilizers

Municipal wastewater, determination of perchlorate in reclaimed, 13, 14*f*

N

National Pollutant Discharge Elimination System (NPDES), 251
New York, Lake Champlain, nutrient over enrichment, 246–247
Nickel
concentrations in phosphate and micronutrient fertilizers, 145*t*
means of trace metals in fertilizers, Lebanon, 98*f*
means of trace metals in N-fertilizers, Lebanon, 93*f*
means of trace metals in NPK-fertilizers, Lebanon, 96*f*
means of trace metals in P-fertilizers, Lebanon, 94*f*
metals in assessment, 128
ranges and means in fertilizers, Lebanon, 98*t*
risk based concentration (RBC) for all scenarios, 140*t*, 141*t*
trace concentrations in K-fertilizers, Lebanon, 95*t*
trace concentrations in N-fertilizers, Lebanon, 92*t*
trace concentrations in NPK-fertilizers, Lebanon, 95*t*
trace concentrations in P-fertilizers, Lebanon, 94*t*
unit RBC value in fertilizer, 143*t*
See also Trace metal content of fertilizers, Lebanon
Nitrate deposits, perchlorate, 33
Nitrate fertilizers
agricultural grade production, 49–50
characteristics of sodium nitrate crystallizers, 50*f*
crystallization temperatures, 50
formulating low perchlorate containing products, 52

general process scheme for low
perchlorate production, 53f
process scheme to increase
production, 52, 55
proposed general process scheme –
ion exchange resin, 56f
proposed sodium nitrate process
scheme, 54f
selective anion exchange resin to
removal perchlorate, 55
separating high and low perchlorate
fractions, 52, 55
technical and refined grade
production, 51
See also Sociedad Química y Minera
S.A. (SQM)
Nitrate products, Chilean, crops, 30
Nitrate salts
fertilizers, 19
sodium and potassium, 20, 22
Nitrification, nitrogen fertilizer, 238,
239f
Nitrogen
concern for excessive fertilization,
209–210
conservation tillage, 221–222, 223t
control of excessive fertilization,
215–216
cycle, 237–240
excessive loading to aquatic
ecosystems, 245
factors affecting runoff and leaching,
162–163
inorganic forms, 162
leaching losses, 163
nutrient criteria, 211, 213
turfgrass, 162
Nitrogen fertilizers
application time, 243
best management practices (BMPs),
241–243
consumption in U.S., 152, 153f
denitrification and leaching, 240
economics, 236
fate during first growing season after
application, 241f

Illinois sales, 236f
immobilization, 240
means of trace metals in Lebanon,
93f
mineralization, 237–238
nitrification, 238, 239f
nitrogen cycle, 237–240
percent of applied fertilizer lost form
tile lines, 240f
proper nitrogen rate, 241
relationship between historical
nitrogen use and quantity of
nitrogen lost, 242f
relative source of corn N uptake,
238f
size of hypoxia zone, 236f
trace metal concentrations in
Lebanon, 91–92
variation in optimum nitrogen rate
over time, 242f
See also Inorganic nutrient use;
Trace metal content of fertilizers,
Lebanon
Nitrogen leaching and runoff
construction of experimental facility,
164
effects of fertilization and
precipitation events on nitrogen,
172–173
facility for accessing, 166f
fertilizer cycles, 167t
layout of mixed-species landscape,
165f
leaching losses per fertilization, 173,
176
maintenance, 165, 167
materials and methods, 164–168
monthly precipitation inputs, 169f
nitrate nitrogen and ammonium
nitrogen in percolate water from
landscapes, 172f
nitrogen contributions from surface
runoff, 168
percolate nitrogen contributions from
rainfall, 168–170
quantity of nitrate nitrogen and

ammonium nitrogen leached daily, 174f, 175f
statistical analysis, 168
volume of percolate measured daily, 170f
water budget summary, 170–172
water sample collection and chemical determination, 167–168
Nitrogen release. *See* Soil incubation methodology
Nitrogen sources, fertilizers, 19–20, 22
Nonpoint source pollution
addressing nutrient enrichment problems, 250
background, 245–247
clean lakes, 249–250
Clean Water Act, 252
coordination of federal programs, 253
Dead Zone of Gulf of Mexico, 246
information and education, 253
Johnson Sauk Lake, 250
Lake Champlain, New York and Vermont, 246–247
Lake LeAquaNa, 250
land treatment implementation, 253
national pollutant discharge elimination system (NPDES), 251
nutrient criteria, 252
nutrient water quality standards, 252
preference for stewardship approach, 251
programs, 247–251
recommendations, 252
restoration and protection projects, 252
Section 314, 249–250
section 319 program, 250–251, 252
total maximum daily loads (TMDLs), 249
water quality management framework, 248f
water quality standards, 248–249
North America. *See* Trace metal content of fertilizers

NPK-fertilizers
means of trace metals in Lebanon, 96f
trace metal concentrations in Lebanon, 95–97
See also Trace metal content of fertilizers, Lebanon
Nutrient losses
concentrations of PO_4-P in runoff solution, 226, 227f
conservation tillage, 221–222
effect of tillage systems on PO_4-P content of runoff, 224f
fertilization application, 222, 225
influence of surface residues and soil moisture conditions, 223t
influence of surface residues fertilizer application timing on cumulative losses, 228t
influence of tillage system on runoff, during rainfall simulation, 223t
loads of PO_4-P in runoff solution, 226, 228f
manure, 230
organic fertilizers, 230
rainfall simulators, 221
soil moisture and tillage system, 225–226, 230
soil test P and runoff P, 230–231
wet versus dry soil, 226, 230
Nutrient management
approaches for developing criteria, 212f
control of phosphorus and nitrogen, 215–216
criteria, 211, 213
high-tech farming, 216
issues in developing programs, 213–214
phosphorus index, 210–211
runoff control, 215–216
See also Excessive fertilization; Site-specific agriculture
Nutrient use
ratio to removal, 157–159
See also Inorganic nutrient use

O

On-farm research, site-specific systems, 201–202
Organic fertilizers, nutrient losses, 230

P

Participant laboratories, perchlorate detection, 26–27
Perchlorate
  absorption and accumulation, 29
  action by IMC Global, 42
  ammonium perchlorate in rocket propellant, 8
  analytical capability by ion chromatography, 39
  analytical method, 38
  analytical program monitoring possible presence in fertilizers, 36–37
  analytical protocols, 37–39
  calibration curves, 38f
  commercial lawn and garden fertilizers, 41, 42t
  common forms, 33
  comparison of AS-11 and AS-16 columns, 37f
  concentrations in commercial fertilizers, 35t
  concentrations in fertilizer components, 34t
  content of Magruder check samples (1993–1999), 40t
  corrected concentrations in fertilizer components, 36t
  detection in sylvite, 28
  determination in Chile saltpeter extract, 12f
  determination in presence of sulfate, 11f
  determination in reclaimed municipal wastewater, 13, 14f
  determination of trace, 9
  determination using AS16 column, 10f
  environment, 16–17
  equipment, 37
  fertilizer source materials by IMC Global, 41
  ground water samples, 9, 11
  health concern, 8, 33
  high ionic strength matrices, 9, 11–12
  interlaboratory corroboration, 24
  ion chromatography as method of choice, 33
  irrigation water as source, 30
  issues for trace analysis of fertilizers, 24
  Magruder check samples, 39
  Method 314.0 for analysis, 9, 13, 15
  method detection limit and calibration, 38–39
  paper entitled Perchlorate Identification in Fertilizers, 34
  potash and langbeinite ore samples, 41t
  reanalysis, 36
  selective anion exchange resin for removal, 55, 56f
  Unregulated Contaminant Monitoring Rule (UCMR) List, 33
  U.S. EPA Contaminant Candidate List (CCL), 8–9
Perchlorate Identification in Fertilizers, paper in 1999, 32, 34
Percolate. *See* Nitrogen leaching and runoff
pH
  acidic deposition mechanism, 101
  influence on aluminum speciation, 104
Phosphate-bearing fertilizer materials trace metal source, 76
  See also Trace metal content of fertilizers
Phosphate fertilizers
  application rates, 115t
  consumption in U.S., 152, 153f

means of trace metals in Lebanon, 94f
metal concentrations, 145 t
trace metal concentrations in Lebanon, 91, 93, 94t
See also Cadmium accumulation; Inorganic nutrient use; Trace metal content of fertilizers, Lebanon
Phosphate sources, fertilizers, 22
Phosphorus
  agricultural nutrient runoff control best management practices (BMPs), 215–216
  comparison of P missed by soil sampling grid sizes, 200f
  concern for excessive fertilization, 209–210
  conservation tillage, 221–222
  control of excessive fertilization, 215–216
  control of P and nitrogen, 215–216
  effect of tillage system on PO_4-P content of runoff, 224f
  high-tech farming and nutrient runoff management, 216
  index, 210–211
  nutrient criteria, 211, 213
  principles of site-specific management, 197
  soil test P and runoff P, 230–231
  total P vs. algal-available P, 210
  See also Excessive fertilization; Nutrient losses; Site-specific agriculture
Plant tissues, measurable perchlorate concentrations, 29
Plant uptake factor (PUF)
  inverse relationship with soil:water partitioning coefficient, 138f
  parameter for risk based concentration (RBC), 134t
  sensitive model parameter, 136–138
Plants, perchlorate uptake, 28–29
Pollution. *See* Nonpoint source pollution

Potash & Phosphate Institute (PPI), 197
Potash (K_2O) fertilizer
  consumption in U.S., 152, 153f
  fertilizer source materials, 41
  trace metal concentrations in Lebanon, 94, 95t
  See also Inorganic nutrient use; Trace metal content of fertilizers, Lebanon
Potassium
  influence of soil test K levels on corn yields, 203, 204f
  principles of site-specific management, 197
  See also Site-specific agriculture
Potassium chloride
  fertilizer source materials, 77t
  intercomparison analyses for KCl, 88t
  sampling schedule, 82t
  trace metal content, 83, 84t
  See also Trace metal content of fertilizers
Potassium nitrate
  agricultural grade production, 49–50
  general process scheme for low perchlorate production, 53f
  proposed general process scheme – ion exchange resin, 56f
  technical and refined grade production, 51
  See also Nitrate fertilizers
Potassium sources, fertilizers, 22–23
Potato
  Cd concentrations, 119–121
  relationship between potato tuber Cd and total soil Cd, 120f
  See also Cadmium accumulation
Precipitation
  effects on nitrogen, 172–173
  monthly, 169f
  percolate nitrogen contributions from, 168–170
  See also Nitrogen leaching and runoff

Precision farming, increasing crop yield, 216
Principles, ion chromatography, 5
Production
  agricultural grade sodium and potassium nitrates, 49–50
  fertilizer, recordkeeping, 23
  increasing, of nitrate fertilizers, 52, 55
  site-specific agriculture, 205
  technical and refined grade potassium nitrate, 51
  See also Nitrate fertilizers
Product safety, risk based concentrations (RBCs) in evaluating, 143–144

R

Rainfall
  effects on nitrogen, 172–173
  monthly, 169f
  percolate nitrogen contributions from, 168–170
  See also Nitrogen leaching and runoff
Rainfall simulators
  nutrient losses, 221
  See also Nutrient losses
Recordkeeping, fertilizer production, 23
Recycled materials, source of trace metals, 125
Reference dose (RfD), parameter for risk based concentration (RBC), 135t
Refined grade potassium nitrate, production, 51
Regulations, ion chromatography, 4
Relative absorption factor (RAF), calculating summary intake factors (SIFs), 133t
Removal/use ratio, nutrient, 157–159
Risk assessment. See Health risk assessment

Risk based concentrations (RBCs)
  application rate (AR), 134t
  conceptual model, 129–130
  equation, 131–132
  evaluating product safety, 143–144
  fraction of land (FOL), 135t
  metals in fertilizers, 138–139
  multi-crop farm scenario, 132
  parameters for calculating RBCs, 134t, 135t
  plant uptake factor (PUF), 134t
  reference dose (RfD), 135t
  setting standards for metals in fertilizers, 143–144
  slope factor (SF), 135t
  soil accumulation factor (SACF), 134t
  soil-water partition coefficient (Kd), 135t
  unit RBC calculation, 132, 136
  unit RBC values for metals in fertilizers, 143t
  upper and lower bound conditions, 140t, 141t
Risk paradigm, scientific thinking and modeling, 125
Runoff
  high-tech farming nutrient management, 216
  See also Nitrogen leaching and runoff; Nutrient losses; Phosphorus

S

Safety, product, risk based concentrations (RBCs) in evaluating, 143–144
St. Augustinegrass
  N leaching or runoff, 176
  turfgrass, 162
Saltpeter, Chile, perchlorate analysis in, 12
Sampling, fertilizers, 25

Section 303, total maximum daily loads (TMDLs), 249
Section 314, clean lakes, 249–250
Section 319 programs, nonpoint source pollution, 250–251, 252
Selenium
  concentrations in phosphate and micronutrient fertilizers, 145t
  digestion techniques, 66t
  instrument techniques, 67t
  metals in assessment, 128
  risk based concentration (RBC) for all scenarios, 140t, 141t
  statistical summary, 70t, 71
  unit RBC value in fertilizer, 143t
  See also Heavy metals in fertilizers
Site-specific agriculture
  comparison of P requirement missed by soil sampling grid sizes, 200f
  crop production, 197
  economic advantages, 201
  economic analysis, 201
  economics of P and K fertilization, 205
  efficiency, 202, 205
  enhancing tools with databases and models, 201
  environmental impact, 202, 205
  focus of 21st century, 206
  Foundation for Agronomic Research (FAR), 197
  grid sampling, 198–199
  grid size, 199
  high yield management, 205
  impact on grain prices, 198
  increasing production, 205
  influence of soil test K values on corn yields, 204f
  management systems, 197–201
  on-farm research, 201–202
  Potash & Phosphate Institute (PPI), 197
  principles for potassium and phosphorus, 197
  relationship between total cost, crop value and level of input, 198f
  separate recommendations for each management zone, 200–201
  soil fertility levels, 202
  soil test K recommendations map, 199f
  soil test summary for North America in 2001, 203f
  top profit producers survey, 198
  world record corn yield, 205
  zone sampling plan, 200–201
Site-specific approach, nutrient criteria development, 214
Slope factor, parameter for risk based concentration, 135t
Slow-release fertilizers
  commercial development, 181
  Controlled Release Fertilization Task Force, 182
  correlation of soil incubation nitrogen release and accelerated lab extraction, 191–192
  data expression, 185–186
  detection, 185
  enhanced efficiency, 181
  extraction equipment, 184–185
  future work, 193–194
  goals of method development process, 183t
  lab apparatus using chromatography columns, 185f
  laboratory method development, 183–186
  nitrogen forms in leachate from Polyon urea, 189f
  nitrogen release over time, 190–191
  nitrogen release vs. incubation days for six materials, 190f
  piecemeal development, 181–182
  principle, 184
  protocol, 185
  release plot of sulfur coated urea (SCU), 186f
  sampling, preparation, and sample size, 184
  soil incubation methodology development, 187–194

See also Soil incubation methodology
Sociedad Química y Minera S.A. (SQM)
  analysis of nitrate ore samples, 48*t*
  caliche nitrate deposits, 47*f*
  commercial use of caliche ore deposits, 46
  company, 46
  nitrate fertilizer products, 45
  sodium nitrate, 19
  use of SQM nitrates in United States, 47–48
  See also Caliche ore
Sodium nitrate
  agricultural grade production, 49–50
  characteristics of, crystallizers, 50*f*
  crystallization temperatures, 50
  general process scheme for low perchlorate production, 53*f*
  nitrogen source, 19–20
  perchlorate analysis in, 12
  proposed general process scheme – ion exchange resin, 56*f*
  proposed process scheme, 54*f*
  See also Nitrate fertilizers
Soil:water partitioning coefficient
  inverse relationship with plant uptake factor (PUF), 138*f*
  sensitive model parameter, 136–137
Soil accumulation factor, parameter for risk based concentration (RBC), 134*t*
Soil incubation methodology
  ammonia volatilization, 187
  ball jar studies, 187
  correlation of soil incubation nitrogen release and accelerated lab extraction, 191–192
  future work, 193–194
  incubation lysimeters, 188–189
  initial studies in plastic bags, 187
  lab data predicting soil replicate, 193*f*
  nitrification, 189–191

  nitrogen forms in leachate from Polyon urea, 189*f*
  nitrogen released over time, 190–191
  organic matter, 187–188
  plot of percent release vs. days incubation for materials, 190*f*
  regression analysis, 192
  relationships between soil release and lab prediction data, 192*f*
  See also Slow-release fertilizers
Soil moisture, tillage system, 225–226, 230
Soil test levels, North America, 202, 203*f*
Speciation, aluminum, 101
SQM. *See* Sociedad Química y Minera S.A. (SQM)
Standards, risk based concentrations (RBCs) in setting, 143–144
Sulfate
  effect on distribution of Al speciation, 105, 106*f*
  perchlorate analysis in presence of, 9, 11
Sulfate of potash magnesium (SPM)
  fertilizer source materials, 77*t*
  sampling schedule, 82*t*
  trace metal content, 83, 84*t*, 85
  See also Trace metal content of fertilizers
Sulfur coated urea, nitrogen release vs. incubation days, 190*f*
Summary intake factors, parameters to calculate, 133*t*
Surveys
  analysis of real world samples, 27–28
  design for heavy metals, 64
  evaluation of participant laboratories, 26–27
  fertilizers, 26–30
  fertilizer screening, 113
  participants for heavy metals, 73–74
  See also Heavy metals in fertilizer
Sylvinite, potassium source, 23
Sylvite

perchlorate detection, 28
potassium source, 23

T

Target cancer risk, summary intake factors, 133t
Target hazard quotient, summary intake factors, 133t
Technical grade potassium nitrate, production, 51
Temperature, nitrification, 238, 239f
The Fertilizer Institute (TFI)
 Controlled Release Fertilizer Task Force, 182
 risk assessment, 125–126
 See also Health risk assessment
Tillage. *See* Conservation tillage
Time, application, nitrogen fertilizer, 243
Tobacco, Chilean nitrate products, 30
Total maximum daily loads (TMDLs), equation, 249
Trace metal, public awareness, 113–114
Trace metal content of fertilizers
 acid reflux digestion technique, 79–80
 analytical wavelengths and method detection limits, 81t
 composite samples for chemical analysis, 78
 di-ammonium phosphate/mono-ammonium phosphate (DAP/MAP) and triple super phosphate (TSP), 82, 83t
 intercomparison analyses for DAP/MAP, 87t
 intercomparison analyses for KCl and AS, 88t
 interlaboratory analytical comparison, 85–86
 interlaboratory analytical comparison method, 80–81
 ion-coupled plasma–emission spectrometers (ICP–AES), 80
 KCl, urea, ammonium nitrate (AN), ammonium sulfate (AS), and sulfate of potash magnesium (SPM), 83, 85
 materials and methods, 76–81
 operating conditions for ICP–emission spectrometer, 80t
 phosphate-bearing fertilizer sources, 76
 sampled fertilizer source materials, 77t
 sample digestion and chemical analysis, 78–80
 sampling rates, 78
 sampling schedule for each fertilizer source material, 81, 82t
 sampling scheme and sample handling, 76–78
 study objectives, 76
 trace metal content of AS and SPM, 84t
 trace metal content of KCl, 84t
 trace metal content of urea and AN, 85t
 wet digestion technique, 79
Trace metal content of fertilizers, Lebanon
 average annual application rates of fertilizers, 96
 comparing possible cumulative additions with tolerance limits of Washington state and Canada, 98t
 K-fertilizers, 94, 95t
 main source of Cd, 96
 materials and methods, 91
 means of trace metal concentrations in 67 fertilizer samples, 98f
 means of trace metals in N-fertilizers, 93f
 means of trace metals in NPK-fertilizers, 96f
 means of trace metals in P-fertilizers, 94f
 N-fertilizers, 91–92

NPK-fertilizers, 95–97
P-fertilizers, 91, 93, 94*t*
ranges and means of trace metal concentrations for 67 samples, 98*t*
trace elements coprecipitating with carbonates, 97
Triple super phosphate (TSP)
  fertilizer source materials, 77*t*
  sampling schedule, 82*t*
  trace metal content, 82, 83*t*
  See also Trace metal content of fertilizers
Turfgrass, nitrogen, 162

U

United States. *See* Inorganic nutrient use
United States Environmental Protection Agency (EPA)
  Method 314.0 for perchlorate, 9, 13, 15
  perchlorate on Contaminant Candidate List (CCL), 8–9
  risk assessment, 125–126
  See also Health risk assessment
Unregulated Contaminant Monitoring Rule (UCMR) List, perchlorate, 33
Urban area, impacting water resources, 162
Urea
  fertilizer source materials, 77*t*
  means of trace metals, Lebanon, 93*f*
  N-fertilizer in Lebanon, 91–92
  sampling schedule, 82*t*
  trace metal content, 83, 85
  See also Trace metal content of fertilizers

V

Vanadium
  concentrations in phosphate and micronutrient fertilizers, 145*t*
  metals in assessment, 128
  risk based concentration (RBC) for all scenarios, 140*t*, 141*t*
Vascular plants, implications of perchlorate, 28–30
Vermont, Lake Champlain, nutrient over enrichment, 246–247
Vollenweider–OECD Eutrophication model, nutrient load to waterbodies, 217

W

Wastewater, determination of perchlorate in reclaimed, 13, 14*f*
Water, irrigation, source of perchlorate, 30
Waterbodies
  evaluating allowable nutrient load, 217
  nutrient-related water quality, 217
  See also Excessive fertilization
Water budget summary
  rainfall, irrigation and percolate, 170–172
  See also Nitrogen leaching and runoff
Water quality
  Clean Water Act, 252
  development and adoption of standards, 252
  excessive fertilization, 208–209
  management framework, 248*f*
  nutrient-related, 217
  standards, 248–249
  See also Nonpoint source pollution
Wheat
  Cd concentrations, 116–117
  fresh weight grain and tuber yields after one year fertilizer application, 119*t*
  grain Cd concentrations over second year application, 118*f*
  Idaho rock phosphate (RP) and grain Cd, 118–119

range of grain Cd concentrations, 117–118
relationship between grain Cd and total soil Cd, 117f
second year of treatment, 117
See also Cadmium accumulation

Z

Zinc
concentrations in phosphate and micronutrient fertilizers, 145t
metals in assessment, 128
risk based concentration (RBC) for all scenarios, 140t, 141t
unit RBC value in fertilizer, 143t

Zinc fertilizers
application rates, 115t
See also Cadmium accumulation

Zone sampling, site-specific systems, 200–201

Bestsellers from ACS Books

The ACS Style Guide: A Manual for Authors and Editors (2nd Edition)
Edited by Janet S. Dodd
470 pp; clothbound ISBN 0-8412-3461-2; paperback ISBN 0-8412-3462-0

Writing the Laboratory Notebook
By Howard M. Kanare
145 pp; clothbound ISBN 0-8412-0906-5; paperback ISBN 0-8412-0933-2

Career Transitions for Chemists
By Dorothy P. Rodmann, Donald D. Bly, Frederick H. Owens, and Anne-Claire Anderson
240 pp; clothbound ISBN 0-8412-3052-8; paperback ISBN 0-8412-3038-2

Chemical Activities (student and teacher editions)
By Christie L. Borgford and Lee R. Summerlin
330 pp; spiralbound ISBN 0-8412-1417-4; teacher edition, ISBN 0-8412-1416-6

Chemical Demonstrations: A Sourcebook for Teachers, Volumes 1 and 2, Second Edition
Volume 1 by Lee R. Summerlin and James L. Ealy, Jr.
198 pp; spiralbound ISBN 0-8412-1481-6
Volume 2 by Lee R. Summerlin, Christie L. Borgford, and Julie B. Ealy
234 pp; spiralbound ISBN 0-8412-1535-9

The Internet: A Guide for Chemists
Edited by Steven M. Bachrach
360 pp; clothbound ISBN 0-8412-3223-7; paperback ISBN 0-8412-3224-5

Laboratory Waste Management: A Guidebook
ACS Task Force on Laboratory Waste Management
250 pp; clothbound ISBN 0-8412-2735-7; paperback ISBN 0-8412-2849-3

Good Laboratory Practice Standards: Applications for Field and Laboratory Studies
Edited by Willa Y. Garner, Maureen S. Barge, and James P. Ussary
571 pp; clothbound ISBN 0-8412-2192-8

For further information contact:
Order Department
Oxford University Press
2001 Evans Road
Cary, NC 27513
Phone: 1-800-445-9714 or 919-677-0977

More Best Sellers from ACS Books

Microwave-Enhanced Chemistry: Fundamentals, Sample Preparation, and Applications
Edited by H. M. (Skip) Kingston and Stephen J. Haswell
800 pp; clothbound ISBN 0-8412-3375-6

Designing Bioactive Molecules: Three-Dimensional Techniques and Applications
Edited by Yvonne Connolly Martin and Peter Willett
352 pp; clothbound ISBN 0-8412-3490-6

Principles of Environmental Toxicology, Second Edition
By Sigmund F. Zakrzewski
352 pp; clothbound ISBN 0-8412-3380-2

Controlled Radical Polymerization
Edited by Krzysztof Matyjaszewski
484 pp; clothbound ISBN 0-8412-3545-7

The Chemistry of Mind-Altering Drugs: History, Pharmacology, and Cultural Context
By Daniel M. Perrine
500 pp; casebound ISBN 0-8412-3253-9

Computational Thermochemistry: Prediction and Estimation of Molecular Thermodynamics
Edited by Karl K. Irikura and David J. Frurip
480 pp; clothbound ISBN 0-8412-3533-3

Organic Coatings for Corrosion Control
Edited by Gordon P. Bierwagen
468 pp; clothbound ISBN 0-8412-3549-X

Polymers in Sensors: Theory and Practice
Edited by Naim Akmal and Arthur M. Usmani
320 pp; clothbound ISBN 0-8412-3550-3

Phytomedicines of Europe: Chemistry and Biological Activity
Edited by Larry D. Lawson and Rudolph Bauer
336 pp; clothbound ISBN 0-8412-3559-7

For further information contact:
Order Department
Oxford University Press
2001 Evans Road
Cary, NC 27513
Phone: 1-800-445-9714 or 919-677-0977

Highlights from ACS Books

Desk Reference of Functional Polymers: Syntheses and Applications
Reza Arshady, Editor
832 pages, clothbound, ISBN 0–8412–3469–8

Chemical Engineering for Chemists
Richard G. Griskey
352 pages, clothbound, ISBN 0–8412–2215–0

Controlled Drug Delivery: Challenges and Strategies
Kinam Park, Editor
720 pages, clothbound, ISBN 0–8412–3470–1

A Practical Guide to Combinatorial Chemistry
Anthony W. Czarnik and Sheila H. DeWitt
462 pages, clothbound, ISBN 0–8412–3485–X

Chiral Separations: Applications and Technology
Satinder Ahuja, Editor
368 pages, clothbound, ISBN 0–8412–3407–8

Molecular Diversity and Combinatorial Chemistry: Libraries and Drug Discovery
Irwin M. Chaiken and Kim D. Janda, Editors
336 pages, clothbound, ISBN 0–8412–3450–7

A Lifetime of Synergy with Theory and Experiment
Andrew Streitwieser, Jr.
320 pages, clothbound, ISBN 0–8412–1836–6

For further information contact:
Order Department
Oxford University Press
2001 Evans Road
Cary, NC 27513
Phone: 1-800-445-9714 or 919-677-0977
Fax: 919-677-1303